Written by Carol Amato and Eric Ladizinsky
Illustrated by Kerry Manwaring
Cover photograph by Tom Nelson

Library of Congress Catalog Card Number: 92-40207

ISBN:1-56565-053-0

10 9 8 7 6 5 4 3 2 1

Lowell House
Juvenile
Los Angeles

Contemporary Books

Chicago

Welcome to the Science Fair!

Congratulations! You've decided to enter a science fair. *50 Nifty Science Fair Projects* can help you! In this book, you'll find 50 step-by-step projects that deal with many areas of science, including botany, biology, perception, optics, magnetism, chemistry, and much more. The ideas have been collected over the years through various classes we've taken at school, experimentation at home, and finally through work in the science laboratory.

There's a lot that goes into putting together a science fair project. Perhaps most important is to start at the beginning—with learning as much as you can about the project you've chosen to do and the scientific ideas behind it. The best way to do that is to research it. Research is where your real work begins. We couldn't include all the information you might need on each topic—after all, there are whole books on each of these subjects!—so you must go to the library and do some research on your own. You can look in encyclopedias, other books on the topic, and sometimes even magazine articles. You can also interview people. Talk to science teachers at school. Gather interesting facts and background information on the topic. If you do a project with a friend, you can "brainstorm" (share ideas) with each other to make your project even better.

The next step is to make your project. (With the exception of the most basic of tools, such as scissors, a ruler, and pencil and paper, all of the materials you'll need are listed in each project.) You can find most materials you'll need in art stores or hardware stores. More technical pieces of equipment can be found at laboratory supply stores (which can be looked up in your local phone book). As for putting your project together, we've provided some basic ideas, but you may think of an even better way! Be creative. If you don't have access to the materials we suggest, think of alternatives that will work. Since you'll be demonstrating your project, make it as visually interesting as possible so fellow fair-goers will be attracted to it. Paint things brightly, and use colored plastic or paper. Most of the projects in this book suggest that you invite viewers to take part in your demonstration, but the more ways you can involve volunteers the better! And during your presentation, be sure to tell your viewers why your project works as it does and what science ideas you're trying to convey.

If you like, you can also display all your research in a colorful, interesting way. Be sure to come up with a clear, informative name for your project. You can create a three-sided stand to present your project (directions are given at the

NOTE: The flask in the upper right-hand corner of each project indicates its level of difficulty, 1 being the easiest and 3 the hardest.

end of this introduction). On bright paper, and with titles large enough to be read from 6 to 8 feet away, write down the purpose of your project, the experiments and tests you performed, what you discovered, and any charts or graphs that will help viewers understand the project.

Our final suggestion is that you practice your project so you feel confident with the information and you're sure everything works the way it is supposed to! Good luck!

Building a three-sided display stand:

PARENTAL SUPERVISION REQUIRED
You can build a three-sided display stand using plywood. Although the one here is a specific size, you can make yours any size you wish.

What You'll Need:

- sheet of plywood 4' x 8' x 1/4"
- piece of wood 2"x 2"x 8' long
- hammer
- finishing nails (no heads)
- any color paint (optional)
- lamp (with a clamp)
- saw
- wood glue

Directions:

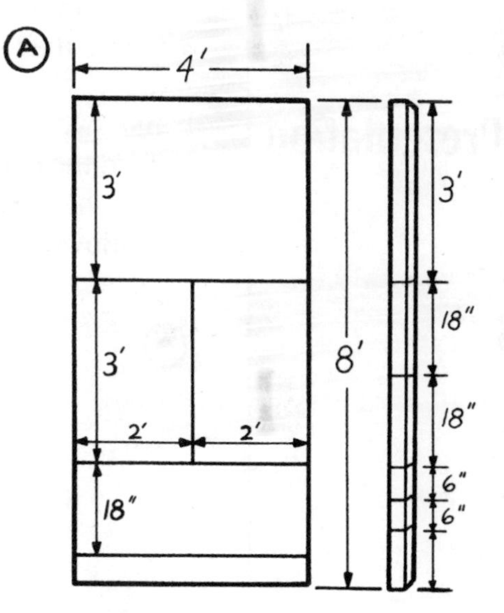

1. With a parent's help, saw the plywood into three pieces, two that are 4' x 3' and one that is 4' x 18". Set one 4' x 3' piece aside for the back of your display and the 4' x 18" piece aside for the shelf (A).
2. Cut the other 4' x 3' piece in half, making two pieces 2' x 3' for the sides of the display.
3. Cut the wood into a piece 3' long, 2 pieces 18" long, and 2 pieces 6" long (A). Discard the remaining piece.
4. Nail the back and sides of the plywood together as shown (B).
5. Glue or nail the short and long wood pieces into place to hold up your shelf. Add the shelf, then paint your stand a bright color. After the paint is dry, attach the lamp.

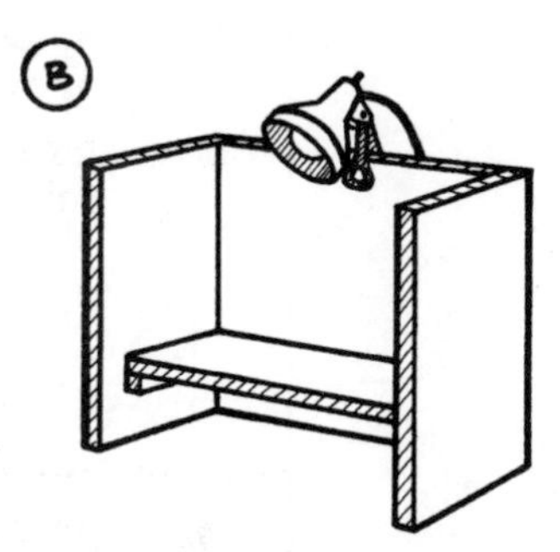

The Shadow Knows!

What color is your shadow? Is it always black? Why can't it be red, or blue, or purple? This project lets you demonstrate that not only can your shadow be a color other than black, but you also can have more than one shadow at a time!

What You'll Need

- a room in which you can turn off the lights
- a large white surface (a wall works best)
- lightbulbs (or floodlamps) in red, green, and blue
- three light sockets (any arrangement that will allow all three lights to focus on the white surface at the same time)

Preparation

1. Put the three bulbs in the sockets, with the green bulb in the middle. Make sure all three bulbs are the same distance from the white surface.
2. Turn on the colored lights. Adjust the angle of the bulbs until the area where they shine is as "white" as possible. Turn them off.

Presentation

1. Ask two volunteers to stand close to the white surface.
2. Turn the colored lights on and the lights in the room off.
3. Ask the volunteers to move around until they each see three colored shadows, then ask them to step back out of the way (A).

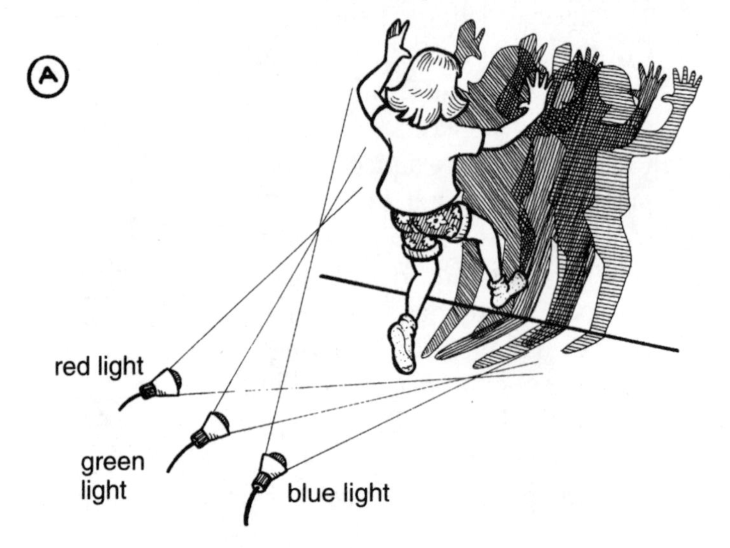

4. Next turn off the red light. What color does the wall show now?
5. Ask your volunteers to again step in front of the wall. What color are their shadows now? Turn the red light on again.
6. Repeat steps 4 and 5 two more times, turning off first the blue light, then the green light. What colors show each time?

7. Ask your volunteers to use their hands as objects in front of the light, rather than their whole bodies. Ask them to move farther away from the wall, then closer again. What do your viewers see?

Why?

Believe it or not, all the thousands—even millions—of shades of colors you're capable of seeing are composed of combinations of just three colors—red, blue, and green. To understand what's happening in your project, stand in profile against a white wall (B). Have a friend turn one colored light on at a time and tell you where the shadow of your profile appears. Notice that when the light to your friend's right is on, your profile produces a shadow to the left. When the light to your friend's left is on, your profile makes a shadow to the right. When the middle light is on, your profile makes a shadow directly behind it. So, your profile takes away (or blocks) different colored light at different places on the wall.

The retina in your eye has three receptors—one for each of the three colors. When all three lights are on, areas of the wall that receive equal amounts of all three colors appear as white. Areas that receive only two colors (because one of the colors is blocked by your hand) are colored. For example, the area where only red and green light mixes looks yellow. You get different combinations of color at different locations, resulting in colored shadows. By knowing which colors are where, try to figure out how to mix the three base colors of light to produce the others.

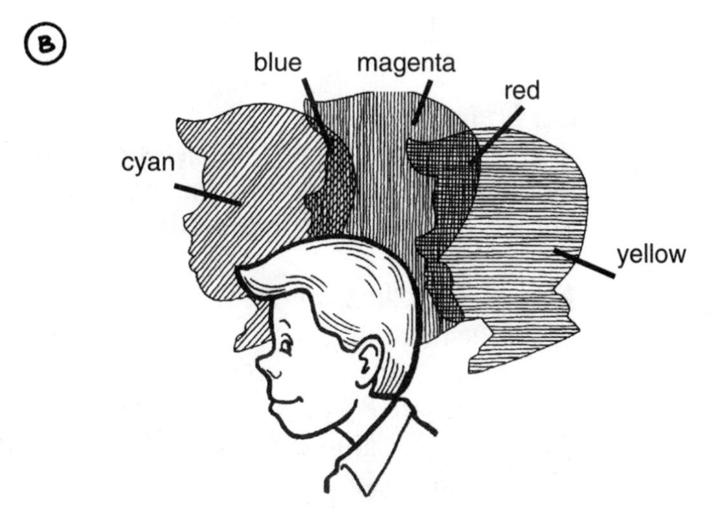

Strong As . . . *Paper?*

Do you think paper is strong? Would you ever think of using paper as a stand to support a stack of books? Paper is not as flimsy as you think! The shape of the stand you make, however, does make a difference, as your viewers will see with this project.

What You'll Need

- six sheets of writing paper
- can
- several books
- cellophane tape
- jar

Preparation

Which shape is the strongest?

1. Fold one sheet of paper into thirds and tape the ends together (A).
2. Using the scissors, cut another sheet of paper in half lengthwise. Fold each half in half again, then open each half so it forms an "L." Put both "L" shapes together to form a square and tape them together as shown (A).
3. Roll a third sheet of paper around the can, overlapping the edges. Tape the edges closed, then remove the can (A).

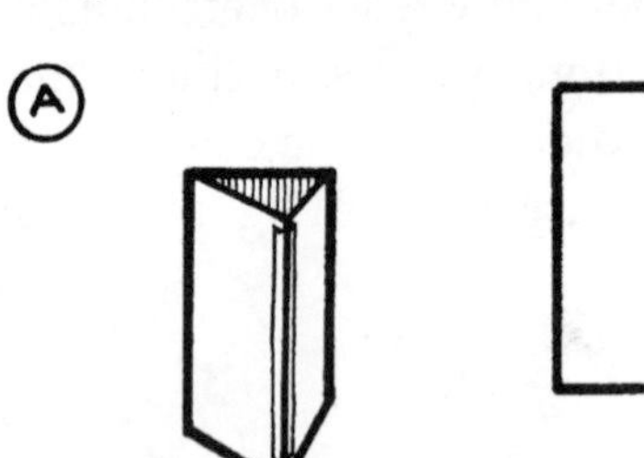

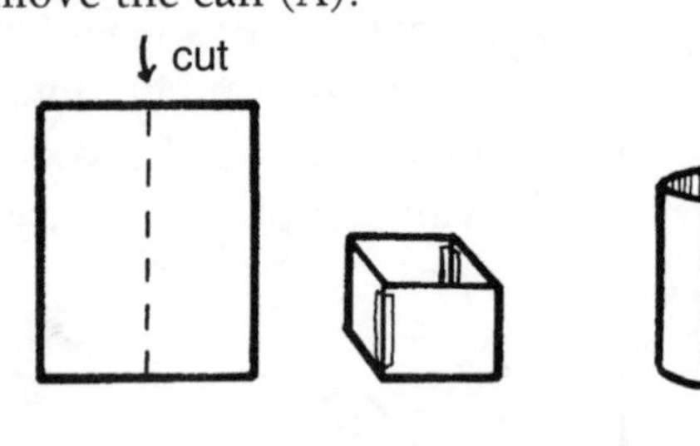

Paper bridge:

1. Fold one sheet of paper into pleats, making each fold about 1/4" wide.
2. Roll two more sheets into cylinders, each with a 3" diameter. Tape the edges of each cylinder together.

Presentation

Which shape is the strongest?

1. Place the three shapes on a table, then ask your viewers which shape they think is the strongest.
2. After your viewers have made their guesses, place one book on each shape. Add books one by one to each shape until it collapses. Which was the strongest?

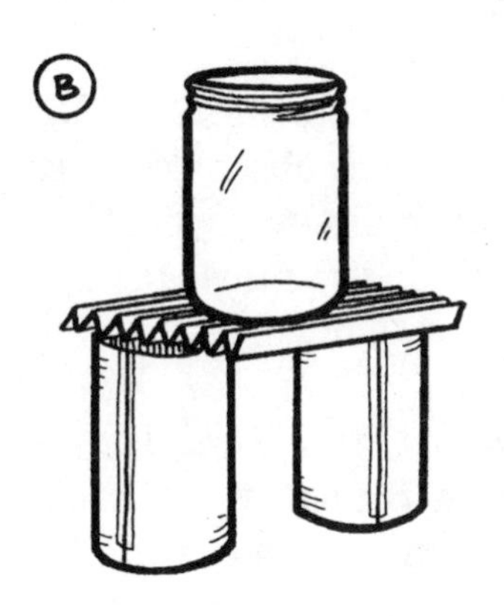

Paper bridge:

1. Stand the two cylinders next to the jar. With the pleated paper hidden from view, ask your viewers if they think the jar can be supported by a piece of paper laid across the top of the cylinders.
2. Then put the pleated paper on top of the cylinders and the jar on top of the pleated paper (B). The pleated paper supports the jar!

Why?

Why is a round cylinder shape so strong? You can understand why by remembering that a chain is only as strong as its weakest link. In the triangle and box shapes, the weight of the books is not distributed (that is, shared) evenly by all parts of the shapes. Instead, the weight tends to be carried more by the folded corners than by the sides. Because of this, the folds reach their breaking point quickly. The hollow cylinder is stronger because the weight of the books is distributed evenly over every part of it (C). Thus, each part always bears the least amount of weight possible, so together the parts can carry more weight.

A flat sheet of paper would never support a jar, but when you fold a sheet into pleats, you add strength to the paper. Why? Because with pleated paper, *more* paper is directly under the jar—enough paper, in fact, to support it. Pleats of paper are called *corrugation*. Corrugated cardboard has pleats inside its outer layers. That's what makes cardboard so strong.

It's Elementary, My Dear

PARENTAL SUPERVISION RECOMMENDED

When police find an unknown substance at the scene of a crime, they ask the police lab to study it. To discover what the substance is, the scientists and criminologists in the lab use *chromatography* to separate the different ingredients. In this project, you will get to be the criminologist as you solve the following crime:

The jewelry store has been robbed. The robber handed the jeweler a note demanding money. The note was written in black ink. The police have arrested four suspects, each of whom has a black pen. To identify the robber and solve the crime, you must analyze the ink in each pen and compare it with the ink on the note.

What You'll Need

- four black ballpoint/marking pens of different brands
- three sheets of filter or chromatography paper, 5" x 5" (available at chemistry supply stores)
- four small white stickers
- wooden stick at least 9" long
- container at least 7" wide and 5" high
- pencil
- pencil holder
- paper towels
- hair dryer
- isopropyl alcohol
- $5\frac{1}{2}$-oz. tin can with the bottom removed
- large jar
- cellophane tape

Preparation

1. Number the stickers 1 through 4 and put a sticker on each pen .
2. On two sheets of the filter paper, draw a pencil line $\frac{3}{4}$" up from the bottom. Then cut the sheets into 8 strips, each strip $1\frac{1}{4}$" by 5" (A).
3. Using the pencil, number each of the strips as follows: two labeled 1, two labeled 2, and so on through 4.
4. With the first ballpoint pen, draw an ink line over the pencil line on both strips labeled 1. With the second pen, do the same on the two strips labeled 2. Do the same for the remaining pens and strips.
5. Tape one set of strips numbered 1 through 4 to the stick, making sure they don't touch each other (B).
6. Fill your container with $\frac{1}{2}$" of water.
7. Put the four numbered pens in the pencil holder, and have the third piece of filter paper ready.
8. Plug in the hair dryer and have it ready for your presentation.

Presentation

1. Ask for four volunteers and hand out a numbered pen to each one.
2. Turning your back to them, ask one volunteer to secretly use his or her black pen to write the following note on the third piece of filter paper: *"Hand over all your jewels! Don't scream! And don't call the police!"* Tell the person to leave a 3/4" margin on each side and to use the same pen to draw a line across the paper about 3/4" up from the bottom (C).

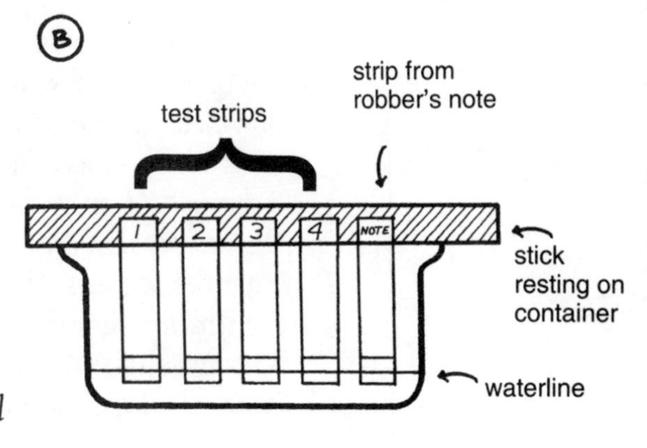

3. Ask the volunteers to put the note in your hand. Then, turning to face the audience, tell them you will discover who the robber is!
4. Cut two 3/4" strips from the sides of the note. Write "NOTE" on the top of each strip (B).
5. Tape one of these strips to the stick with the other strips, making sure this strip doesn't touch the others.
6. Rest the stick across the top of the container so that the strips are just touching the water. The ink lines should be above the water. Let sit for 10 minutes. Use this time to explain that you are demonstrating the science of chromatography, and that criminologists use the tools of chromatography to solve crimes.

7. After 10 minutes, check the strips. The ink on some strips should have separated into different colors. Carefully take them off the stick, lay them flat on paper towels, and dry them with the hair dryer.

8. Show the strips to your audience, and write down the colors that show up on each strip. Which one has the same pattern on the note the bank robber wrote? Which pen was it? *NOTE: Are there some strips in which the ink did not separate? Scientists use other chemicals to test these. You will use the alcohol. Explain this to your audience. Follow steps 9 through 11 only if the ink on one or more of your strips did not separate. Otherwise, skip to step 12.*
9. Tape the second set of strips in the same order around the outside of the tin can, making sure they do not touch each other and that their bottoms align with the bottom of the can.
10. Pour $^1/_2$" of isopropyl alcohol into the jar, then put the can with the test strips in the jar. Repeat steps 7 and 8. While you are waiting, explain that you are further testing the strips with the alcohol. Also, from your research, tell your viewers about some of the different tests police scientists use.
11. After the 10 minutes are up and the strips are dry, write down the colors that appear. Compare them to your robbery note test strips. Which one has the same pattern as the note the bank robber wrote? Which pen was it?
12. Have your volunteers hold up their pens so you can see the numbers. Tell the person whose number matches, "You're the robber!"

Why?

The ink in each pen is actually a mixture of different substances (chemical compounds), each of which may be very different or slightly different in color. Each of the different chemicals is attracted to water, alcohol, or the filter paper to a different degree. How strongly one chemical is attracted to (and bonds with) another chemical is called its *affinity*. When the paper strips are dipped in water, the water slowly creeps up tiny spaces in the filter paper due to capillary action (see Drink Up!, page 35). As the water passes through the ink line, chemicals in the ink that have a strong affinity for the water get carried with the upward-moving water. Chemicals with a smaller affinity for water (or a larger affinity for the paper) either get left behind or travel upward more slowly. In this way, the different chemicals in the ink get separated from each other. Inks made of the same chemicals will show the same pattern of separations. If the ink chemicals have little or no affinity for water, another liquid like the alcohol can be tried. Because the separated chemicals are usually different colors, the technique is called *chromatography* (in Greek, *chromo* means "color").

This same idea is used to separate chemicals for producing medicines and a whole host of commercial chemical products.

Listen Up!

We all know that sound travels through air, but does it travel through water? Or through a solid material? With this project, you'll be able to "listen" for the answers to these questions!

What You'll Need

- funnel
- rubber tubing, 18" long
- string, 20' long
- two tin cans
- large washtub
- two metal pipes, each about 2" in diameter and 2' long
- hammer and nail
- wristwatch

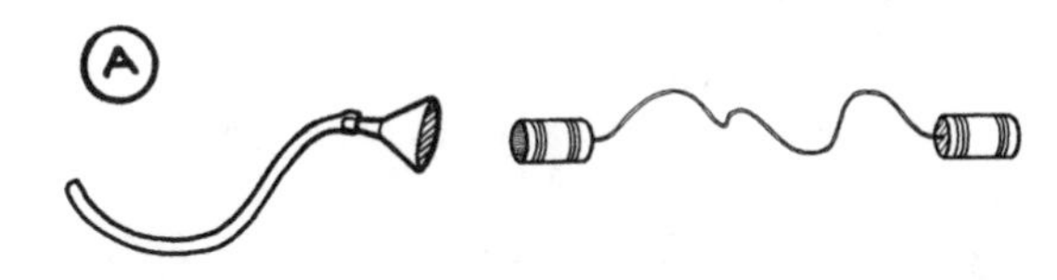

Preparation

Start your project by creating two tools for hearing sound (A).

Stethoscope: Put the pointed end of the funnel into one end of the rubber tubing.

1. First, using the hammer and nail, punch a small hole in the bottom of each tin can.
2. Then put either end of the string through the hole in each can and tie a knot in the ends so they won't slip out.

Presentation

Sound traveling through water:
(Prior to the demonstration, fill the washtub with water.)

1. Give the stethoscope to someone in your audience. Ask him or her to put the funnel part all the way in the tub of water and the end of the hose to his or her ear (B). *NOTE: You can also ask the person to put his or her ear to the surface of the water.*
2. Then stick one metal rod in the water and tap on it with the second metal rod (B). What does the person hear?

Sound traveling through a solid material:
Last, demonstrate that sound can travel through a solid material—namely, the string in your telephone.

1. Have a viewer hold one end of the string telephone while you hold the other end. Move away from each other until the string is tight (you shouldn't be able to hear each other talk).
2. Now talk into the tin can. Ask your volunteer a question. Can you hear her answer? Yes!

(C) MODEL OF A LIQUID

A sound vibration travels through a material, such as water, creating compression regions.

compression region

MODEL OF A SOLID

A sound vibration travels very quickly through a solid because the molecules are close together.

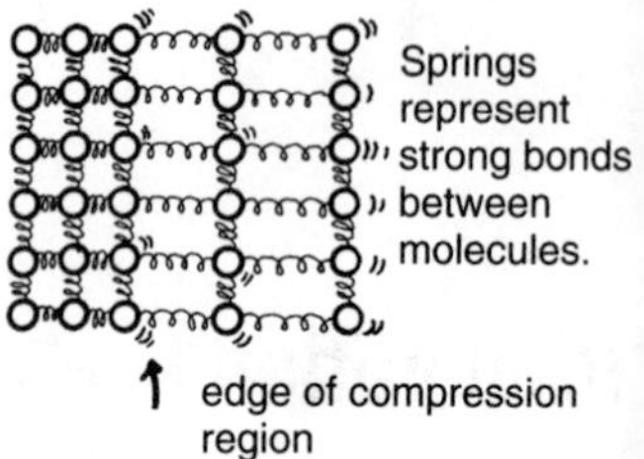

Why?

When you strike an object, you cause the molecules in it to vibrate. As they vibrate, they push on the molecules next to them, creating a *compression* region of tightly packed molecules. The packed molecules push on the molecules next to them, causing the compression to move forward. In this way, a vibration travels from one compression region to the next, to the next, like falling dominoes. Eventually, the vibration reaches your ear as sound (C). When a sound vibration travels through a narrow space (as through the stethoscope), the molecules causing the vibration collide with the side walls of the hose. This prevents the vibration from spreading outward and weakening. As a result, the sound transmits better and so is louder.

Through which material do you think sound travels the fastest—air or water? The *stronger* the molecules are connected to each other in a material, the better a vibration can travel from one molecule to the next. The *closer* the molecules are packed together, the less time it takes a vibration to travel between them. So, sound vibrations travel faster and farther through water than through air because the molecules in water are bonded more tightly together. Sound vibrations travel fastest through a solid, such as the telephone string.

Can you imagine why putting your ear to a train track would help you know a train is coming? A train's sound vibrations travel faster through the metal track than they do through the air!

And They're Off!

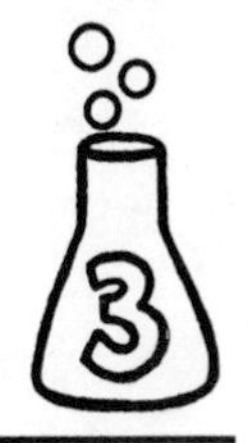

Why is it that things that are in motion—like a ball tossed into the air or a car driving on a highway—eventually come to a stop? Why don't they just continue moving forever? The answer is friction. In this project, you can demonstrate how different surfaces create different amounts of friction.

What You'll Need

- several toy trucks (Matchbox™ or Hot Wheels™ size)
- piece of plywood, 2' x 3'
- tape measure
- small paper flags of different colors
- enough books to make two stacks 10" high
- 2'-long strips of rug, linoleum, wax paper, aluminum foil, Plexiglas™, cloth, fine sandpaper, and rough sandpaper (only five items needed)
- glue
- yardstick
- marking pen

Preparation

1. With a marking pen, draw four lines down the plywood to create five lanes. The lanes should each be $7\frac{1}{5}$" wide (A).
2. Glue some flags along the lane lines and across the back of the board.

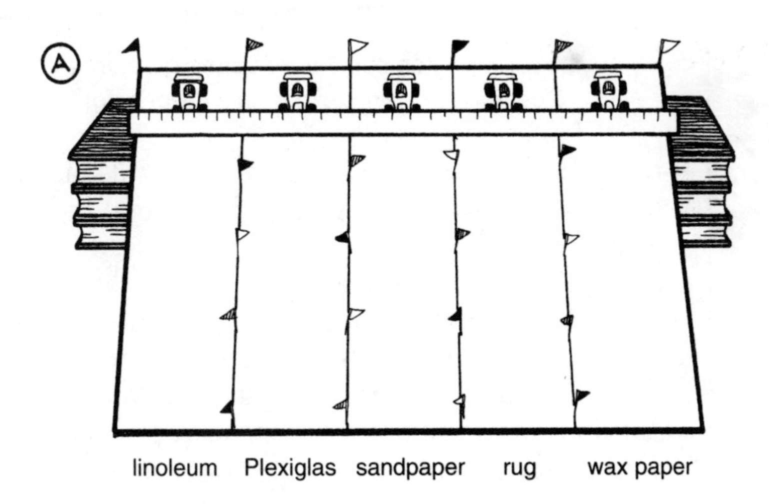

3. Now lean the plywood against two 5"-high stacks of books to make a ramp as shown.
4. Place the strips of different material (any five you choose) under the ramp, one at the base of each lane. Make sure the strips stick out at least 20" from the ramp.

Presentation

1. Ask five people from the audience to volunteer to be race car drivers. Give each person a car and a lane, and have each predict which car will roll the farthest on the different surfaces.
2. When everyone is ready, hold the yardstick in place as shown so everyone can place their cars down (A).
3. Now lift up the yardstick and let 'em roll! Which car won?
4. Use the rest of the books to raise both ramps another 5", then repeat step 2. Did the increased height help any of the cars?

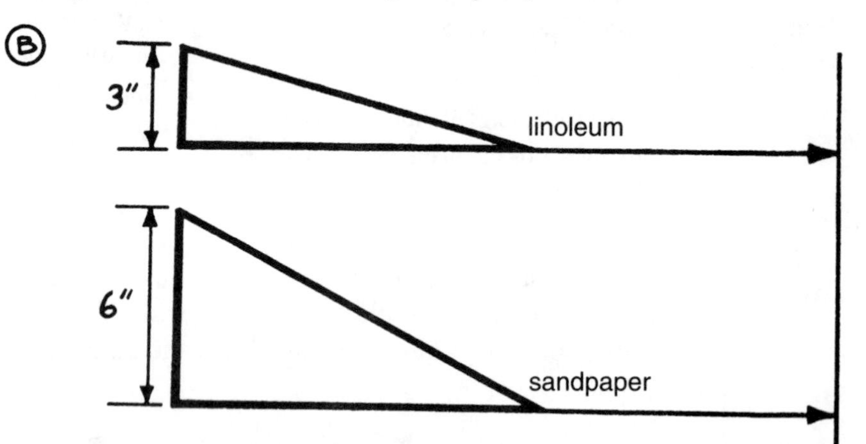

A car rolling on sandpaper from a greater height can roll as far as a car rolling on linoleum from a lower height.

Why?

No surface is truly smooth, not even glass. If you had a powerful enough microscope, it would show tiny hills, valleys, and cracks on even the most polished and shiny of surfaces. When two surfaces are rubbed together, these hills and valleys get caught on each other, and it requires energy to break them apart. The larger the hills on the surfaces, the more they stick together and the rougher the material feels to the touch.

The resistance you feel when rubbing two surfaces across each other is called *friction*. In your project, as the cars roll down the ramp, their tires must constantly break free from surface bumps. That breaking free requires energy, so the cars slow down. The rougher the surface, the harder to break free and the slower the car rolls. Without a constant supply of new energy (stepping on the gas), a real car would eventually lose all its motional energy to friction and come to a stop.

What happened when you raised the ramp higher? All the cars rolled farther. That's because, starting higher, they could gain more motional energy as they traveled down the steeper ramp (B).

Can you imagine what would happen to a moving object if there were no friction? In the emptiness of space, where there is no friction, an object, once moving, will glide along until the end of time!

Freefall!

PARENTAL SUPERVISION RECOMMENDED

When you push the "down" button in an elevator at the top of a tall building, you feel yourself falling. But after a split second, the sensation is gone, and you feel as though you're just standing there. Why is that? Let's see.

What You'll Need

- Plexiglas™ tube, with lid, at least 20" long and 8" wide
- plastic astronaut (or other human figure), 4" to 5" tall
- piece of string, 2' long
- ladder
- large pillow
- red paint

Preparation

1. Make a small hole in the end of the tube (the store at which you buy the Plexiglas will likely be able to drill one for you).
2. Tie the astronaut to one end of the string. Pull the other end through the hole in the Plexiglas and tie a knot in it as shown. Make sure the knot is large enough so the string doesn't fall back inside the tube (A).
3. Paint the word "NASA" down the tube using red paint (optional).
4. Put the tube on a table, stand the ladder next to the table, and put the pillow on the floor in front of the ladder.

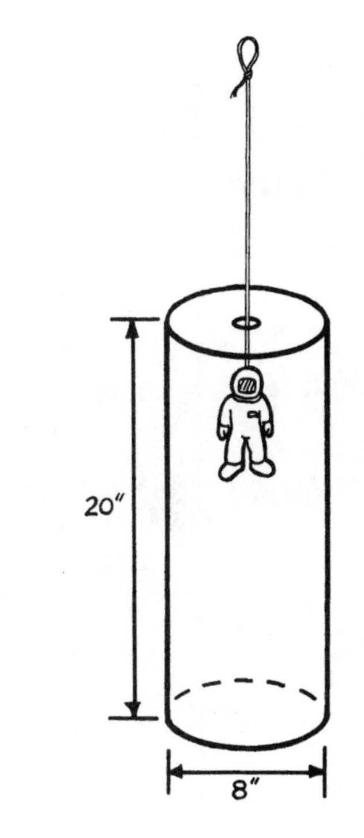

Presentation

1. With everyone watching, pull the string through the lid so the astronaut hits the lid.
2. Let the string go. The astronaut falls to the bottom of the tube. Tell your viewers that you have just demonstrated the force of gravity pulling the astronaut back to earth.

3. Now ask an assistant to hold the ladder while you climb it, bringing the tube with you. *NOTE: You may use a taller ladder than the one pictured here. If you do, be sure to have an adult hold it while you climb.*
4. Pull the string again so the astronaut is up against the lid.
5. Asking your viewers to watch the astronaut carefully, drop the tube and astronaut together onto the pillow. The tube will hit the floor first, *then* the astronaut (B). Freefall!

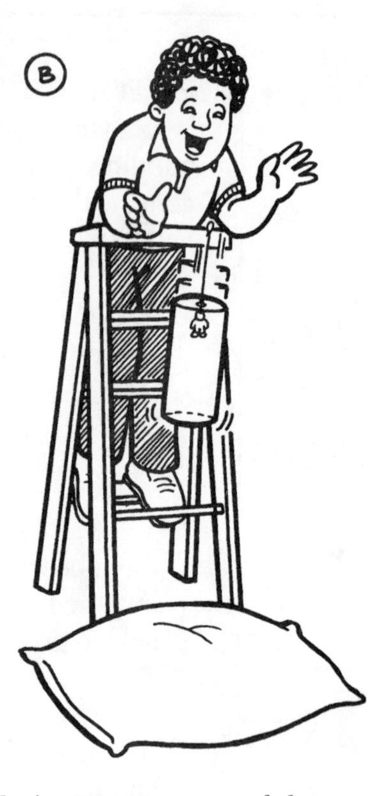

Why?

To understand how you can float in a spaceship (and why the astronaut in your project floated in air), conduct this experiment using only your imagination. Picture yourself in an imaginary elevator 10,000 miles high. You're jumping up and down going crazy, waiting to reach the 5 millionth floor. Just as you're jumping around, the elevator cable snaps and you and the elevator begin to fall to the basement 10,000 miles below. But you don't feel the fall! Why? Since everything falls in the earth's gravity at the same speed, the elevator floor will fall away *from you* as fast as you are falling *toward it*. For every single foot you fall, the elevator floor falls, too.

Now suppose the *basement* were also falling away. You'd never hit bottom and never touch the floor. You'd float forever! But this is just a dream—could it ever really be so? Well, if you were an astronaut in orbit around the earth, you and your spaceship would fall together toward it (like the elevator toward the basement). If you were traveling fast enough, the earth's surface would curve away from under you as fast as you were falling toward it. So "forever falling" is not crazy or imaginary—it's simply an orbit around the earth!

A Breath of Fresh Water

Did you know that plants "breathe"? They not only breathe in carbon dioxide (CO_2) and breathe out oxygen (O_2)—just the opposite of you and me—but they breathe out water, too. But do all plants breathe out the same amount of water? Let's see.

What You'll Need

- a large stalk of a succulent plant, such as aloe or kalanchoe (available at a plant nursery)
- modeling clay
- a bright lamp
- four glass jars
- a large, thick plant leaf complete with stem
- two pieces of cardboard, 6" x 6"

Preparation

The night before your presentation:

1. Cut a hole in the center of a piece of cardboard and put the leaf stem through. Then, using the clay, plug up the hole on both sides.
2. Fill one jar about halfway with water. Put the cardboard over the top so that the leaf stem sticks into the water.
3. Place the second jar upside down over the leaf. The clay will stop any water from escaping into the upper jar from the lower one.
4. Now repeat steps 1 through 3, this time with the succulent stalk.

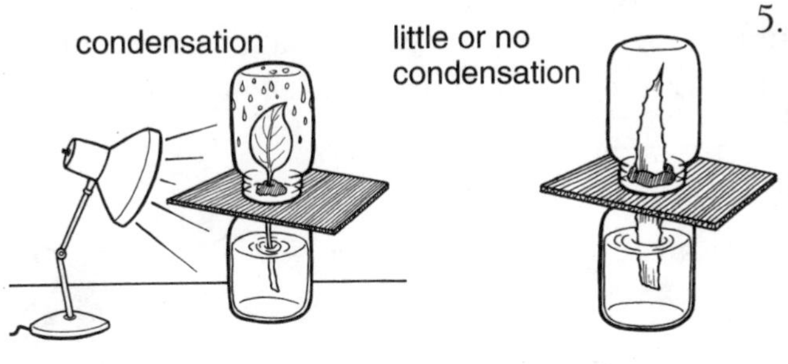

5. Put both sets of jars under the lamp. What happens and how long does it take? Droplets of water form on the inside of the upper jar with the leaf, but not with the succulent stalk.

Presentation

Display the two sets of jars on a table. Explain to your viewers why the set with the leaf has water droplets while the other does not.

Why?

Have you ever dried your hands with a blow drier? When you spread the water over a larger area, or make the air hotter, or blow the air faster, the water disappears faster. In plants, water travels up through the stem to the leaves. There, some of the water is lost by evaporation to the air. Since water is essential to a plant's survival, plants in hot, dry, windy places have evolved to minimize the area on their leaves, and they have developed special surfaces to keep water from escaping. That's why the succulent "breathed out" very little or no water. Being a plant of the desert, it can't afford to lose precious water!

Shape Up!

Why does water take on the shapes it does? For instance, why are raindrops round instead of square? It turns out there are special forces that exist at the surface of water (or any liquid) that give it a particular shape. These forces result in *surface tension*. With a little creative shape making, you can show surface tension at work.

What You'll Need

- modeling clay
- glue
- straws
- yarn (or twist ties)
- pipe cleaners
- dishpan
- 10 cups of water
- $^1/_2$ cup of dishwashing liquid
- $^1/_2$ cup of corn syrup (or granulated sugar)
- toothpicks
- water

Preparation

1. In the dishpan, mix together the water, dishwashing liquid, and corn syrup. Make sure this mixture has no bubbles on the surface.
2. Make different shapes out of toothpicks by dipping the toothpicks in glue, then joining them together with pieces of modeling clay (A). Use your creativity to come up with some interesting shapes besides the ones shown here.
3. Cut pieces from the straws and join them together with yarn (B).
4. Put pipe cleaners into straws to make bubble holders. Create any shapes you like, even three-dimensional ones (C).

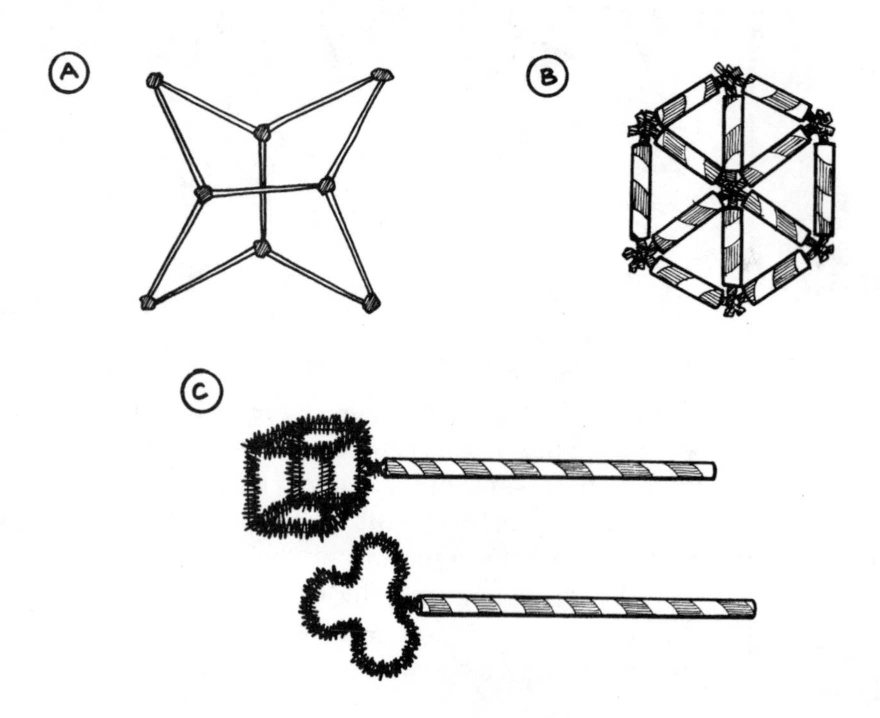

Presentation

1. Dip the toothpick and straw shapes into the soapy water. How many bubble surfaces form in the shapes' sections?
2. Dip a bubble holder in the soapy water and blow a bubble. What happens to the shape of the bubble? What happens when you blow a bubble through a different-shaped bubble blower?

Why?

The molecules that make up a liquid, unlike those that make up a solid, can roll and slide around each other, causing the liquid to change its shape. You can understand why a blob of liquid molds itself into special shapes by thinking about pulling on a wagon. If the wagon is pulled equally from all directions, it doesn't move. Nothing changes. However, if it is pulled more from one side than from another, it will move (D). And it will continue to move until all the forces on it (all the pulls) balance out.

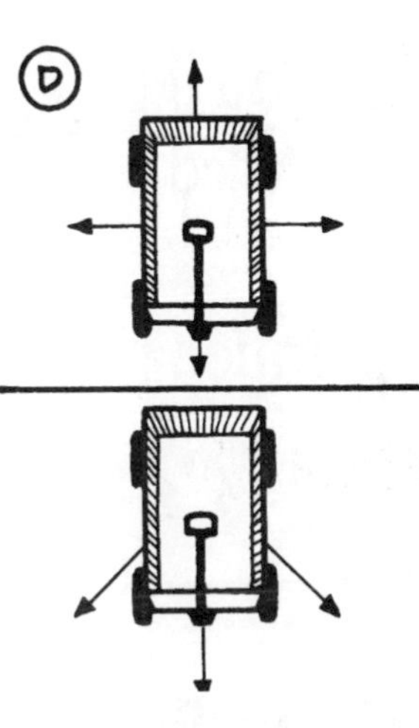

A wagon pulled more in one direction than another will move.

Look at the diagram of the liquid blob (E). A molecule inside the blob gets pulled from all directions equally by the molecules around it, but a molecule on the blob's edge receives a pull only from one side—*into* the blob. As a result, the edge molecules move inward as far as possible, changing the shape of the blob until they can't squeeze inside any further. This is what happened in your project. When you blew irregularly shaped bubbles, the molecules on the surface of the bubbles wanted to move into as small a shape as possible, so they moved into a sphere! With the shapes you made out of toothpicks and straws, why did just one bubble form in each section? Because of surface tension, the bubbles pulled themselves into the smallest units possible: a single shape in each section.

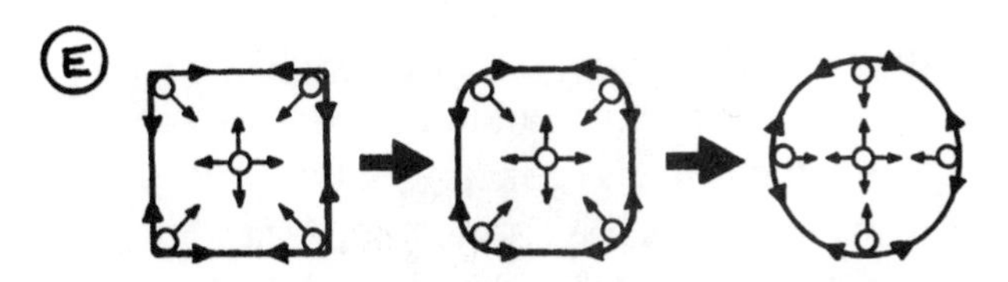

The surface tension on a cube-shaped blob of liquid will pull the liquid into a sphere (the circles are individual molecules).

Tissue Issue

A flimsy piece of paper tissue would rip if you punched a wooden dowel into it, wouldn't it? Maybe not! With this project, you can demonstrate just how strong paper really can be!

What You'll Need

- paper towel tube
- a paper tissue (such as Kleenex™)
- rubber band
- salt
- dowel, 1"to $1\frac{1}{2}$" in diameter (or a broomstick)

Preparation

1. Place the paper tissue over the end of the paper towel tube and use the rubber band to secure it as shown (A).
2. Pour salt into the tube until it is 3" to 4" deep.

Presentation

1. Tell your viewers that the paper towel tube you're holding contains salt, then ask someone to volunteer to rip the tissue with the dowel.
2. Hand the volunteer the paper towel tube and the dowel. Ask him to poke the dowel into the open end of the tube and try to break the tissue (B). He'll never succeed!

Why?

Salt is made up of crystals. When the dowel hits the salt, the downward force is diverted by the multisided crystals in different directions, pushing the salt crystals here and there in a random way. Thus, the force is broken up into many sideways directions, too. As a result, the paper tissue receives just a fraction of the original force from the dowel—not enough to tear it.

Any Which Sway

PARENTAL SUPERVISION RECOMMENDED
With this project, you can dramatically demonstrate the concept of *natural frequency.*

What You'll Need

- six Plexiglas™ rods, all 1/4" in diameter, two 1' long, two 1 1/2' long, and two 2' long
- six hard, brightly colored plastic balls, three 1 1/2" in diameter, three 2 1/2" in diameter
- carpenter's glue
- piece of wood, 2" x 4" x 2'
- hammer
- drill

Preparation

1. With a parent's help, drill six 1/4" holes in the wood.
2. Drop some glue into each hole, then tap a rod into each hole using the hammer. The rods should be in the order shown.
3. Next, drill a hole in each plastic ball. Put some glue in the holes and put one ball on the end of each rod, alternating the large and small balls.

Presentation

With your viewers watching, slide the wood back and forth on a table. The rods will start swinging—some violently, some hardly at all. *NOTE: Don't let them vibrate too wildly—the rods may break!*

Why?

Think back to the last time you pushed someone on a swing. Do you remember that you pushed on the swing as it moved away from you, not as it came toward you? By matching your pushes to the natural movement of the swing, you added energy to the swinging, and the swing went higher. Without realizing it, you were matching the swing's *natural frequency*.

What determines how fast or slow a natural frequency is? Well, with swings, the shorter the chains are, the faster the swing moves back and forth. But the bigger the person sitting in it is, the slower it swings. In your project, the balls on the rods are like kids on swings with chains of different lengths. When you shake the wooden board, little vibrations push on the rods. When these pushes match the natural frequency of a rod, it swings wildly. The long rods swing naturally slower than the short ones. When two are the same height, the one with the bigger ball on it swings more slowly.

A *Cube-Shaped* Globe?

PARENTAL SUPERVISION RECOMMENDED
If the earth were cube-shaped instead of round, how would life be different? With this project, you can see!

What You'll Need

- soft hollow rubber ball, about 6" in diameter
- small squeeze bottle
- empty half-gallon milk carton
- two thumbtacks
- blue food coloring
- modeling clay
- two dowels (or unsharpened pencils)
- plastic tablecloth

Preparation

1. With a parent's help, cut the ball in half to create a hemisphere, and cut a cube shape from the bottom of the milk carton.
2. Now put a thumbtack through the top of one hemisphere, then into the end of a dowel. Do the same with the milk carton cube (A).
3. Now stand the dowels in lumps of clay as shown.
4. Fill the squeeze bottle with water and add blue food coloring.
5. Put the cube and the hemisphere on the plastic tablecloth.

Presentation

1. Spin the half ball while a volunteer squirts the center with a steady stream of blue water. What pattern does the water make on the ball? How much of the liquid reached the bottom edge of the ball?
2. Repeat with the cube (B). Does the same fluid pattern appear on the cube? The bottom edge of the cube?

Why?

Imagine that the blue water in your project represents the earth's atmosphere. You can see by the way the blue water swirled to all areas on the half ball that a spinning sphere shape allows for coverage of its entire surface. Because of the shape of the spinning earth, its atmosphere can flow to all places on the globe. On the cube shape, however, the blue water flicks off the top surface, leaving most of the cube's surfaces untouched. Chance are great that life couldn't exist—or even develop—on a cube-shaped earth.

"Seeing" Is Believing

PARENTAL SUPERVISION RECOMMENDED

Have you ever looked at something really bright, and then when you looked away, you still could see an image of that object? That image is called an *afterimage*. In this project, you can let your viewers "see" what we mean.

What You'll Need

- flashlight
- white paper
- large cardboard box
- small cardboard box (small enough to fit in the large box)
- two cardboard squares, 2" x 2"
- cellophane tape
- stopwatch (or watch with a second hand)
- blank wall

Preparation

1. Cut a piece of white paper big enough to cover the flashlight lens. Tape the paper over the lens.
2. Next cover the white paper with strips of opaque tape, but leave part of the paper uncovered so the light can shine through. You can make this uncovered part an interesting shape, such as a triangle or star (A).

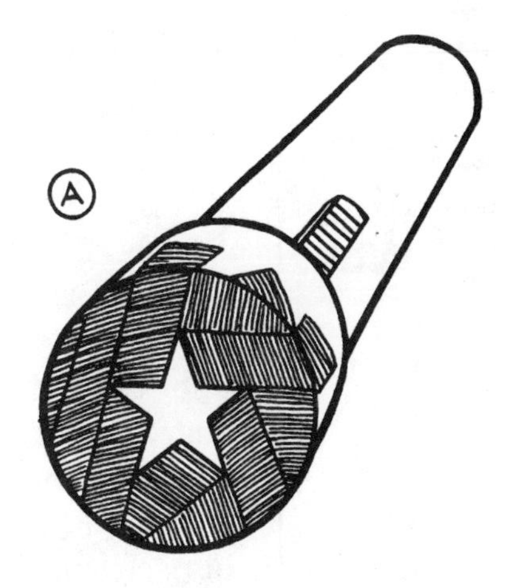

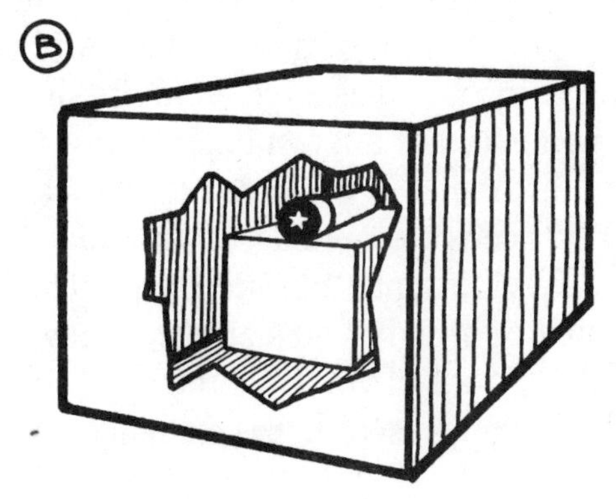

The flashlight rests on a small box inside a large box.

3. Put the small cardboard box inside the larger one, then lay the flashlight on top of the small box (B).
4. Turn the flashlight on, mark an X on the large box where the light hits it, then turn the flashlight off.

5. Cut two 1" peepholes near the X. Make the holes close enough together so that when you put your face up to the box, you can see inside (C).
6. Then tape two cardboard "trapdoors" over the peepholes.
7. Set up your "afterimage" box on a table so that the peepholes face your viewers.

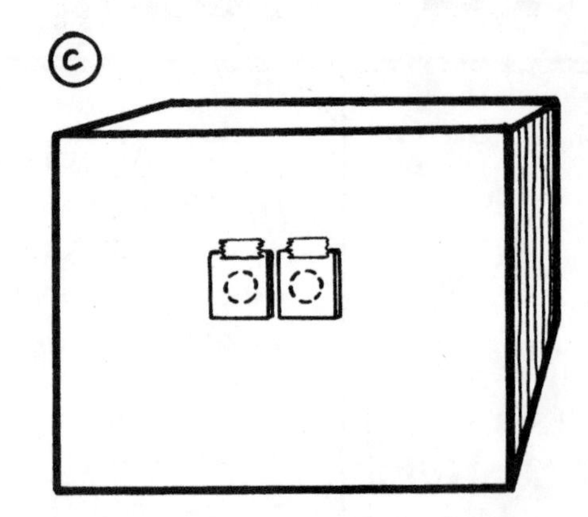

peepholes underneath "trapdoors"

Presentation

1. Turn the flashlight on, making sure the light shines in the direction of the peepholes.
2. Ask a volunteer to look through the peepholes and stare at the brightly lit shape for about 50 seconds.
3. Then tell the person to look at a blank wall and describe what they see. They will describe the shape you made with the opaque tape—an afterimage.

Why?

"Seeing" is not simply taking a picture of what's really around you. What you "see" is a combination of a few things: light reflected from (or given off by) the objects around you; images your brain expects to see; and traces of things your eyes have just been exposed to seeing. By staring at a bright image for a long time, you change the way your eyes function immediately after you stop looking at the image. Light entering your eye hits the light-sensitive lining (the retina) at the back of the eye. The light causes changes in the chemicals that transmit images to your brain. When part of your retina is exposed to bright light long enough, that part becomes less sensitive to light. The retina now has a region on its surface in the shape of the bright light you stared at. For a few moments, this region is unable to send bright images to the brain. So when you then look at a bright white wall, the affected region sends a dim image of the wall to your brain, whereas the unaffected region sends a bright image. Thus, you "see" a dark area in the shape of the bright light you stared at—an afterimage. You're not seeing what's really in front of you (just the wall). Instead, you're seeing a past image (the shape) recorded as a change in the condition of your retina.

Over the Rainbow

If you walk on the beach and look at the waves, you can learn something about the secrets of light. In this project, you'll be able to see waves of light that are very similar to ocean waves!

What You'll Need

- two plates of glass from picture frames, about 8" x 10" (Plexiglas™ will work well, too)
- black construction paper
- desk lamp

Preparation

Place the lamp on a table and plug it in. Have the glass plates and construction paper ready.

Presentation

1. While viewers watch, place the black construction paper on the table, then put two glass plates back to back on the paper.
2. Now shine the lamp light on the glass. What do your viewers see? Notice all the wavy patterns. Where do they come from?
3. Ask a volunteer to press on a corner of the glass plates. How do the wave patterns change? Then ask them to lift the plates and squeeze the plates together. The pattern will change.

Why?

Think about ocean waves. When two bumps on the water (wave crests) combine, they join to form a bigger bump. If a bump collides with a dip in the water (a wave trough), the bump falls into the dip and the water looks flat—as if no waves were there at all. Now imagine a long train of ripples coming toward the shore. Since the shoreline is uneven, parts of the ripples will hit the shoreline first (where the shore sticks out most). Other parts of the ripples will hit later. As the ripples bounce off the shore and travel back outward, they combine, some bump to bump, some dip to bump.

The same thing is happening in your project. Between the glass plates is a thin gap of air. Some of the light falling on the glass bounces off; some of the light passes through the top plate, crosses the air gap, and bounces off the bottom plate. The bouncing light waves combine. Where the waves add together (bump to bump), you see light. Where they cancel each other (bump to dip), you see dark lines.

Chaotic Pendulum

Could the flapping of a butterfly's wings in China drastically affect the weather in New York City? Build a unique pendulum that helps you explore the answer.

What You'll Need

- lightweight string or thread
- wooden dowel, about 16" long and as thick as a pencil
- two large bookends (or wooden sheets), 12" high
- masking tape
- seven donut-shaped ceramic magnets (available at electronic supply stores)
- sheet of wood, about 16" long and wide enough to support the bookends
- glue
- ruler

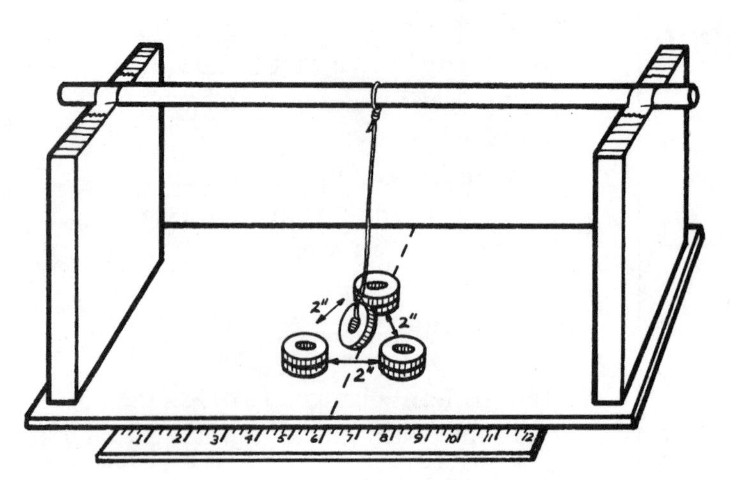

Preparation

1. Stack all seven magnets in a single stack. This will align their poles—all the norths up and all the souths down (or vice-versa).
2. Take the magnets off the stack, two at a time, making sure not to flip them over. Make a triangle of magnet pairs, 2" between each pair, in the center of the wood sheet as shown (use the ruler as a guide). Glue them in place.
3. Assemble your pendulum as shown, taping the dowel to the bookends and hanging the string from the dowel. Adjust the string length to make sure the pendulum magnet barely misses hitting the stacked magnets as it passes over them.

Presentation

1. Pull the wood sheet with the magnets on it out from under the bookends. (You'll experiment with a regular pendulum first.)
2. Pull the pendulum toward you so it's over the 6" mark on the ruler and let it swing until it stops. Have your viewers observe its motion. Now pull the pendulum toward you again, but just a little to the right, say, at the 6 $^{1}/_{4}$" mark. Let go. The pendulum's motion should be almost the same.
3. Now place the wood with the magnetic triangle under the pendulum again, and center it as before. Repeat step 2. The motion will be very complex and unpredictable. You should notice that a very slight change in the initial position of the pendulum results in a totally different pattern.

Why?

Without the magnets in place, the pendulum swings very regularly and predictably. Most important, a small change in the pendulum's starting position produces only a small change in how it ends up.

Most of the physical systems scientists can study are like this. For instance, the orbits of planets and moons are very regular and predictable. Small influences we can't measure don't cause large changes in how these bodies move. That's why astronauts can travel to the moon and be certain it will be there! With magnets, the situation is much more complicated. The way the three magnet stacks affect the pendulum depends on where the pendulum is and how it's moving. The slightest change in how it started can produce big changes in how it ends up. Since we can never know exactly how the pendulum starts (its position, speed, and direction), we can't know how it will move.

One system with this type of behavior is the weather. Since we can't know exactly how the air is moving *everywhere* on earth—how every butterfly is flapping its wings and affecting the air—we can never really predict the weather with total accuracy. The flapping butterfly in China could produce huge effects over time. Systems that behave unpredictably like this are called *chaotic*.

Fighting the Air Force

PARENTAL SUPERVISION RECOMMENDED

By moving a little air around you can create forces more powerful than the mightiest team of horses!

What You'll Need

- two identical plastic mixing bowls with wide, flat rims and a plastic ring at the bottom (used to stand bowls up)
- beeswax
- nylon cording (or shoelaces)
- surgical tubing, 4" long
- large spring clip
- drill
- "sticky" bubble gum

Preparation

1. Drill a hole in the side of one of the bowls large enough for the surgical tubing to fit through. Push the tubing in about an inch (A).
2. Chew up the sticky bubble gum until it is gooey and can stick to anything, then squish it around the base of the tube to seal any gaps. Depending on the type of bowl, if the gum doesn't seal well, try rubber cement or something similar.
3. Drill three holes, equally spaced, around the plastic ring at the bottom of each bowl.
4. Thread a piece of nylon cording through each hole, knotting the ends so the cording doesn't pull through the hole. Tie the ends together as shown to create handles (A).
5. Spread a thin layer of beeswax on the flat rims of both bowls, making sure to completely cover the rims.

A

holes drilled into plastic ring

nylon cords tied to form handle

sticky bubble gum seals hole

plastic tubing

layer of beeswax

Presentation

1. Tell your viewers that, using the simple tools before you, you're going to make a device so strong that even the strongest of volunteers won't be able to pull it apart.

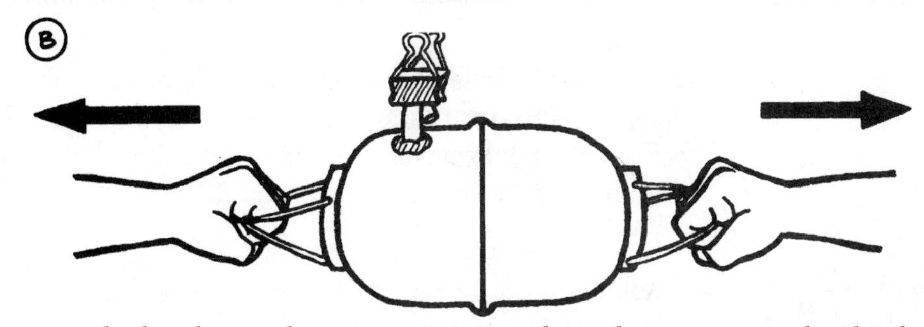

2. Put the bowls together rim to rim, matching them up precisely. Slowly press them together so a good seal is made.
3. Now put your mouth over the tube and suck out as much air as possible. Then, with the tube still in your mouth, pinch and bend the tube and clamp it with the spring clip to keep air from getting back in.
4. Have two volunteers grab opposite handles and try to evenly pull the bowls apart (B). If you've done it right, they will find it difficult, if not impossible, to separate the bowls. After their attempt, remove the spring clip and see how easily the bowls separate!

Why?

All around you are molecules of air that are constantly flying around, colliding with each other and everything else. The constant colliding of air molecules with the sides of the bowls creates a pushing force. When you first put the bowls together, there are as many molecules pounding on the inside surface of the bowls (pushing out) as there are molecules pounding on the outside (pushing in).

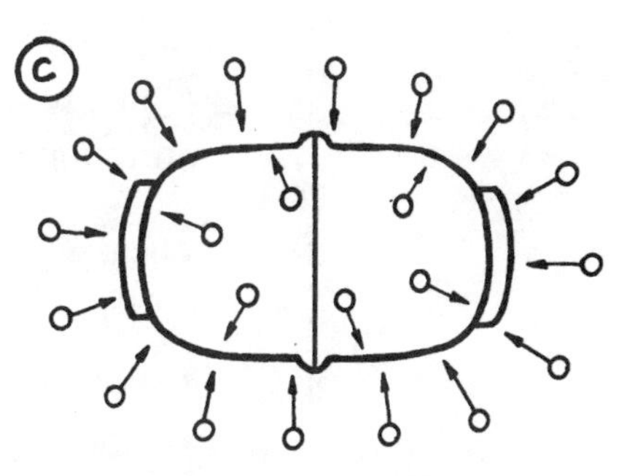

Fewer molecules bombard the walls from the inside of the bowls than from the outside.

When you sucked air out from the inside, though, there were now fewer air molecules inside to push outward. As a result, the push inward was now much stronger than the push outward (C). The millions of collisions on the large surfaces of the bowls combine to create a huge force that makes it nearly impossible to separate the bowls.

A similar experiment was once performed in which two teams of horses tried to pull apart two large metal hemispheres—and failed!

All Thumbs

What would life be like if we didn't have our thumbs? Would we be able to do all the amazing things we do—like playing baseball, or creating works of art, or flying an airplane? When you think about it, we owe a lot to our thumbs, and this project shows you just why.

What You'll Need

- Velcro™
- pair of pants with button at waist
- shoes with shoestrings
- pencil, pen, and paper
- several different-sized coins
- hairbrush and comb
- objects of different sizes and shapes (like chopsticks, a paperweight, a battery, a book, and tape)

Preparation

Arrange all your items on a table.

Presentation

1. Ask a volunteer to hold out a hand and move his thumb around in different directions. Then ask the person to move his fingers around like that. Your volunteer won't be able to do so.
2. Now invite the volunteer to do the following tasks with both thumb and fingers:
   - button and then unbutton the pair of pants
   - tie the shoelaces on the shoes
   - pick up the coins
   - write his name on the sheet of paper
   - pick up the flashlight, then turn it on
   - brush and comb his hair
   - try to use any other object you collected
3. Your volunteer, of course, will be able to easily accomplish these things. But now loosely pin your helper's thumbs to the palms of his hands using the Velcro™ (you don't want to cut off the circulation) (A). *NOTE: You can also sew down the thumb on a glove and use that.*
4. Ask the volunteer to repeat all the tasks (B). He'll be "all thumbs"!

Why?

Only primates (humans, apes, and monkeys) have hands that can grasp and hold objects. That's because primates have an *opposable* thumb—it moves opposite to the other fingers (C). And among the primates, humans can move their thumbs the farthest across their hands. Why is this so important? Without our thumbs, human civilization wouldn't have evolved the way it has. Because of this special physical feature, the earliest humans could do many things that other animals couldn't, such as making fires, cutting stones into tools, and making bows and arrows. Each time a new tool was invented, the range of human experience was broadened, and that led to developmental advances in our brains. With advancing intelligence, more and more complex tools could be invented, which led to more and more brain development. Eventually, humans were even making computers and spaceships!

River of Heat

PARENTAL SUPERVISION RECOMMENDED
When metal touches a heat source, the heat moves through the metal until the metal becomes hot, too. This process is called *conduction.* Do all metals conduct heat at the same rate?

What You'll Need

- candle and candle holder
- iron wire, 2' long
- aluminum wire, 2' long
- piece of wooden board, 2' x 1'
- watch with second hand
- 12 paper clips
- oven mitts

Preparation

1. Twist both ends of the iron and aluminum wires together and lay them down on the board (A).
2. Line up six paper clips on each wire as shown.
3. Ask a parent to light the candle, then drop wax on the paper clips to seal their tips to the wires.
4. Blow out the candle and put it in the holder for your presentation.

Presentation

1. Ask your viewers if they think iron and aluminum conduct heat at the same rate.
2. While you put on the oven mitts (*very important!*), once again have an adult light the candle for you.
3. With your viewers watching, hold one twisted end of the wires over the candle flame (B). How long does it take each clip to melt off? Which is the better conductor, aluminum or iron?

Why?

Every material is made up of tiny atoms, each of which is surrounded by negatively charged electrons. "Heat" is actually the combined vibrations and movements of the atoms and these electrons. The more violently the atoms vibrate and the electrons move, the hotter a material is.

When you heat one end of the wires, heat energy from the candle causes the atoms and electrons there to vibrate. Electrons move more easily and faster through the aluminum wire than through the iron wire. So, the aluminum wire heats up and melts the wax first.

Here's Looking at You, Kid

PARENTAL SUPERVISION RECOMMENDED
Do you think it's possible to see your brain making mistakes? Go face to face with your face and find out!

What You'll Need

- glue (or tape)
- bright light source
- large mixing bowl
- spatula
- talcum powder and powder puff
- a shallow cardboard box (larger than your head)
- black construction paper (as big as the box)
- plaster of paris
- wax paper

Preparation

1. Line the inside of the box with wax paper.
2. Following the instructions on the plaster of paris box, mix the plaster and water in the bowl. Make the plaster just thick enough to hold a shape. You'll need enough plaster to fill the box.

3. Pour the plaster mixture into the box until it's almost full. Smooth the surface flat with the spatula.
4. Now powder your face with a thick layer of talcum powder (to keep the plaster from sticking to your face).
5. With your eyes closed, press your face into the mixture as shown (A). Then carefully pull away from the box. If you look into the box now, you'll see an exact impression of your face. Let the plaster harden overnight.
6. Now make a border for the impression of your face. Take a piece of black construction paper and cut out a large oval in the middle. Place the paper over the plaster so that the impression is exposed but the rest of the plaster is covered.
7. Set the box on a table or shelf with the face upright and at eye level (B). Focus the bright light on it so the face is thoroughly lit.

Presentation

Ask a volunteer to stand facing your plaster face. Ask her to look at it as she walks to the right and to tell your viewers what she sees. The face will appear to turn and follow her! Have her walk to the left, bob up and down, run right and left. No matter what she does, the face will seem to follow, turn, bob, and nod. Have another viewer try to escape the face's gaze. Then let everyone look closely and touch the inverted impression.

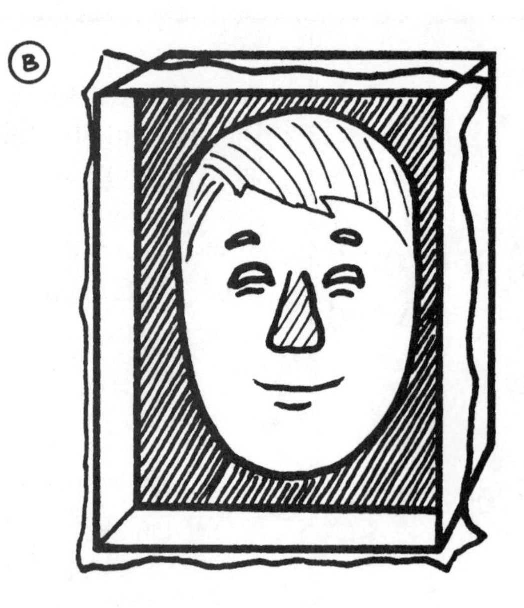

Why?

How is it the face seemed to be able to respond—to move as the volunteer moves? The key to understanding this effect is to use your nose!

Look at the nose on the plaster face from a distance. Is it sunken in or sticking out? You see it sticking out even though you know it's sunken in. In fact, the whole face looks as if it's sticking out. Even though you *know* the face is a sunken impression, your brain still perceives it as a regular face.

Remember "Seeing" Is Believing (page 23)? What you see depends a lot on what patterns your brain is used to seeing. You almost never see an inverted face, so your brain reconstructs the plaster image to match a very familiar pattern—a regular face (on which the nose sticks out).

As you move around a regular face, part of it disappears from view—namely, the side furthest away from you, which is blocked by the side nearest you. With the inverted face, no part of the face sticks out to block your view of it, so you see the whole face as you move. Since the only way you see an entire regular face as you move is if the face turns with you, your brain constructs the familiar pattern of a moving face instead of an inverted one. As a result, the face appears to be following you!

19 Drink Up!

Have you ever wondered how plants "drink" water all the way up to their top leaves? In this project, you can colorfully show how.

What You'll Need

- 12 tall drinking glasses
- 12 white carnations
- glass pipettes (available at a laboratory supply store)
- blue, red, and green food coloring
- water

Preparation

1. A few nights before your presentation, fill three drinking glasses with water. Add three drops of blue food coloring to one glass, three drops of red food coloring to another glass, and three drops of green food coloring to the third.
2. Put a carnation in each glass (snip off the ends first). The next day, you'll see that the color has risen to different levels in the carnations. These levels match the different speeds at which the carnations can draw up water over about 12 hours.
3. For your demonstration, prepare more glasses with carnations for different periods of time—3 hours, 6 hours, and 9 hours, for example.
4. Set all your carnation glasses on a table in order of number of hours each flower has sat in a colored liquid. Display the number of hours on cards.

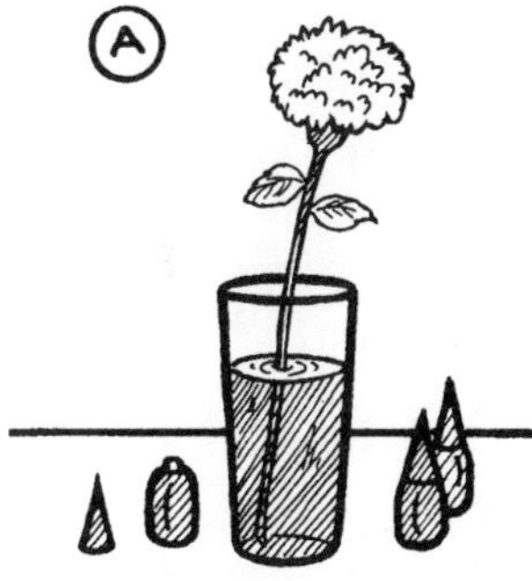

Presentation

1. Explain capillary action to your viewers and show them your display of carnations. Your viewers will see that, over time, colored water rises to the top of the carnations.
2. Hand pipettes out to volunteers. Ask them to each place a pipette in a glass of colored water and let them watch capillary action as it happens.

Why?

To understand how water can travel upward, fill a plastic jug with water, make a hole in the cap, and put a straw halfway into the jug through the hole. Now squeeze the jug's sides. See how the water rises into the straw? The same thing happens in the *capillaries* (tiny tubes) in a plant. When you put a carnation in water, some of the water has to move to make room for the stem. The entire surface of the water doesn't move up, because surface tension is constantly pulling the surface down (see Shape Up! page 18). So, the only place for the water to go is *into* the stem. Water has a lot of surface tension, so it moves all the way up the carnation.

The Big Crash

A huge number of things in the world—from a rocket blasting off to the inner workings of our sun—can be understood in terms of little "billiard balls" (atoms) colliding into each other. Scientists have invented special ideas to describe just how atoms collide. The most important of these ideas is called *momentum*.

What You'll Need

- Plexiglas™ tube, 2' long and at least 2¼" in diameter
- two dowels, 3' long
- eight plastic or wooden balls, 1" in diameter
- square of acetate, 3" x 3"
- plastic or wooden ball, 2" in diameter
- books or other object about 6" thick
- plastic rod, about 8" long
- different colors of model paint
- paintbrush

Preparation

1. Have your local hardware or plastics store cut four holes along one side of your tube as shown. The holes should be large enough for the plastic rod to fit through (A).
2. To make the balls really stand out, paint stripes or spirals on them, using a different color for each ball.
3. Line up the two dowels on a table, leaving about ¾" between them. Tape them to the table surface (B).
4. Put the piece of acetate over one end of the dowels.
5. Rest one end of the tube on the acetate as shown, and rest the other end on a stack of books.
6. Put the plastic rod through the top hole in the tube, and drop a small ball into the top end. The rod will keep it from rolling (A).
7. Set four balls on the dowels as shown (B).

Presentation

1. Ask your viewers to predict what will happen when you pull out the rod so that the ball rolls down and hits the other balls. Will the balls on the dowels move? Will the ball from the tube stop rolling on impact, or keep moving?
2. Repeat step 1 for the other three holes, again asking your viewers to predict what will happen. Notice that the rolling ball does not move as fast, so the balls on the dowels do not move away as fast. In addition, the balls on the dowels move away a shorter distance each time.

3. Let the large ball roll from each of the holes. Ask your viewers to guess what will happen now. Notice that the large ball hits the balls on the dowels harder than a small ball does, and that this force is greatest when the ball is dropped from the topmost hole.

Why?

When watching balls smashing into each other, what ideas help you predict what will happen? A large heavy ball released from the top of the tube causes a bigger effect on the target than a small, lighter ball. So is it how heavy the ball is (its mass) that allows you to predict the effect? No, because a small light ball moving fast (dropped from the very top hole) can cause as large an effect as a large heavy ball moving slowly (dropped from the bottom hole). It takes *two* ideas—mass (how heavy an object is) and speed (how fast it is traveling)—to predict the effect of the collision.

Scientists came up with the idea of multiplying mass and speed and calling this number *momentum*. When you know an object's momentum (*both* its mass and speed), then you can predict the outcome of a "collision course." In your project, the more momentum a rolling ball had, the more motional energy it transferred to the stationary balls. That's why they moved when struck!

Beyond Explanation

PARENTAL SUPERVISION RECOMMENDED

The most important and exciting thing for scientists to discover is something they don't understand. That's because they are then forced to question their old ideas and come up with new and better ways of thinking. In this simple project, you'll see an effect that required a revolution in our ideas about the world.

What You'll Need

- large cardboard sheet
- tape
- polarizing plastic sheet, at least 98% efficient, 7" x 24" (available at laboratory supply stores)
- bright light source

Preparation

1. Cut out three 7" x 7" squares from the polarizing plastic.
2. Frame each plastic square with cardboard. Attach little cardboard feet to the frames so they can stand upright (A).
3. Set two of the framed plastic sheets on the table, about 5" apart and parallel to each other.

Presentation

1. Have one volunteer hold a bright light behind the first polarizer while a second volunteer views through the other polarizer.
2. Have the viewing volunteer grab the frame of the second polarizer in front of her and slowly rotate it (keeping it parallel and close to the first polarizer). She will see the flashlight behind the other polarizer get brighter and dimmer as she rotates the plastic. For one position of the polarizer, no light from the flashlight will get through at all, and the plastic will look black. Have the volunteer hold the second polarizer in this position.

3. Now insert the last filter between the other two. The viewing volunteer will see light shine through! (If not, rotate the middle plastic until she does.) The two polarizers blocked all the light, but adding a third polarizer allowed the light to shine through again. How is it possible?

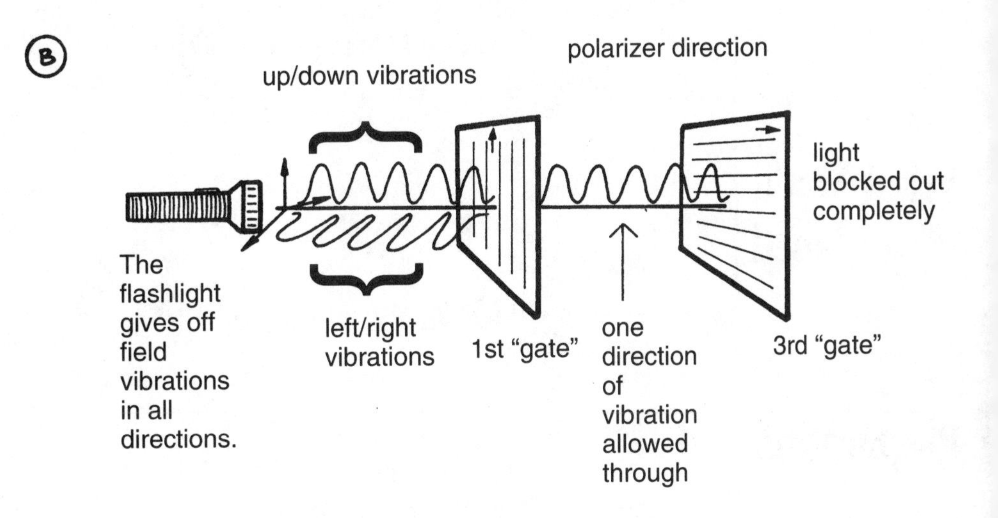

Why?

When a string vibrates, it can vibrate in many different directions—up and down, left and right, or at some angle in between. Electric and magnetic fields can also vibrate (or fluctuate) in different directions. For example, an electric field fluctuating up and down will move charged particles up and down.

Light is a combination of fluctuating electric and magnetic fields. The light from the flashlight is a complicated mixture of field vibrations in different directions (up/down, left/right, etc.) (B). A polarizer works by acting like a gate that only allows one kind of vibration to pass through. In your setup, the first gate (polarizer) is positioned to allow up/down vibrations through, and the second is rotated to allow only left/right vibrations through. So the up/down vibrations from the flashlight pass through the first filter, but can't pass through the second filter. That's why no light shines through.

Understanding why adding a third filter allows light through requires a completely new way of thinking about nature. This new thinking is called *quantum theory*, and it is stranger than the weirdest science fiction! Without quantum theory, we wouldn't have transistors or lasers.

Hot! Hot! Hot!

PARENTAL SUPERVISION RECOMMENDED
This project challenges you to fit a regular screw through a screw eye. But here's the catch: The screw eye is too small! Can you somehow change its size without ruining it? Now imagine this: The screw is the key to get into a castle, but the keyhole is too small. This project will show you how to open the door!

What You'll Need

- screw eye
- glass of ice water (with ice cubes)
- large-head wood screw (the head should be just a bit too large to fit through the screw eye)
- two thick wooden dowels, 8" long
- mercury thermometer
- screwdriver
- a heat source, such as a hot plate, alcohol lamp, or candle

Preparation

1. Screw the large-head screw into the end of one dowel and the screw eye into the end of the other dowel (A).
2. Set the heat source and glass of ice water on a tabletop.

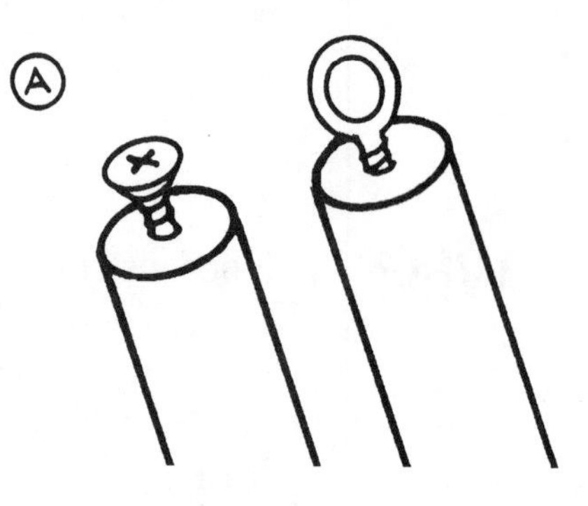

Presentation

When the screw eye is cold, it's too small for the screw head to fit through.

1. Show your viewers that the screw head is too large to fit through the screw eye (B). Tell them you have a way of making the screw eye big enough to fit.
2. Now hold the screw eye over the heat source for a few moments. When you try to put the screw through now, it'll slip right through!
3. Now dip the screw eye (with the screw still through it) into the ice water. When you're sure they're cooled off, ask one of your viewers to try pulling the screw out. He won't be able to!

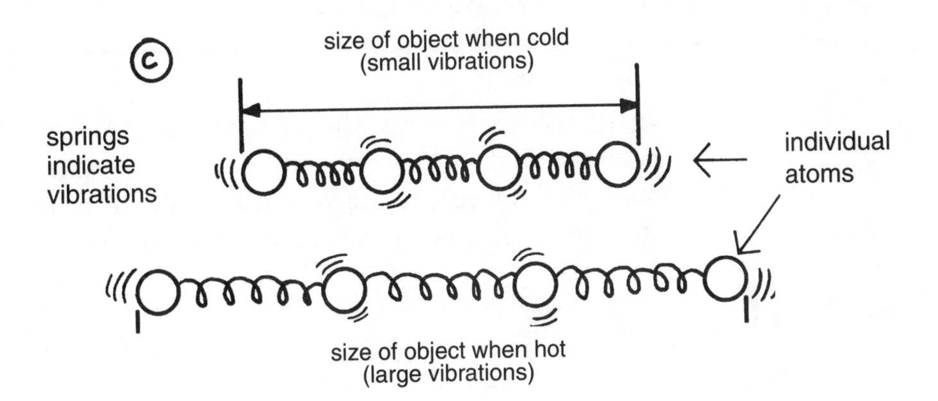

Why?

Even though the metal screw eye seems rigid and still, it's actually made up of tiny atoms that are constantly jiggling around. You can imagine the screw eye as being made up of little balls connected by springs (C). The balls represent metal atoms, and the springs represent the forces holding the atoms together. The little balls in the screw eye are constantly vibrating back and forth. The screw eye's *temperature* is a measure of how large these vibrations are. As the screw eye is heated, the vibrations get larger and larger. This causes the screw eye to grow larger and larger, because the atoms need more and more space to vibrate (D). When the screw eye is hot enough, the hole in it becomes wide enough for the wood screw to pass through.

THREE-DIMENSIONAL OBJECT

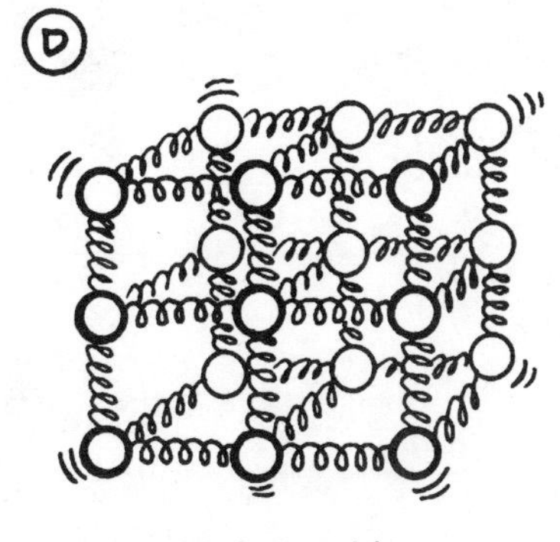

size when cold

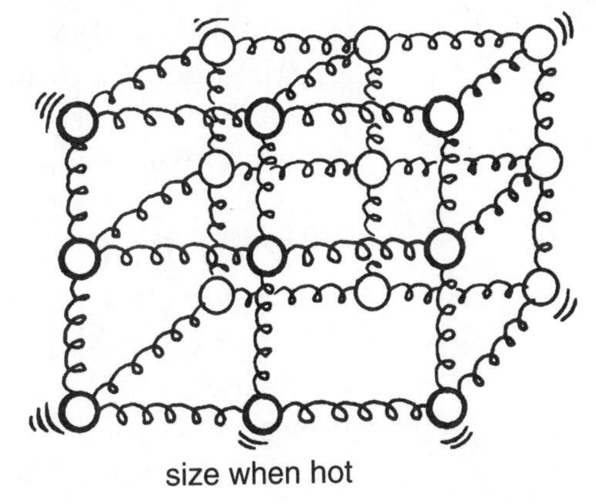

size when hot

Hellooooo!

When two people are standing apart, they can't hear each other very well. Could they hear each other better if an object were placed between them? You don't think so? Amaze your friends by showing that *focusing* sound can make a soft whisper sound like a shout!

What You'll Need

- large balloon, 3' to 4' in diameter
- carbon dioxide (CO_2) tank (look in your phone book for a supplier of compressed gases)
- high-pitched buzzer
- round washtub, 2' in diameter

Preparation

Just before your demonstration, fill the balloon with CO_2 gas and set it on the washtub as shown (A).

Presentation

1. Ask a volunteer to stand on one side of the balloon while you stand on the opposite side. Turn on the buzzer.
2. Ask the volunteer to move around on her side of the balloon until she finds the place where she hears the buzzer the loudest.
3. Then remove the balloon and turn on the buzzer again. Ask the volunteer to say how loud the sound is now.
4. Replace the balloon and buzz the buzzer again, only this time hold the buzzer higher up. Where is the loudest point now?

Why?

Sound vibrations are very much like ripples expanding on a pond. The buzzer in your project sends out ripples of sounds (sound waves). The ripples spread out over a larger and larger area. As they do so, they get weaker and weaker and finally die out (and can no longer be heard). The CO_2 molecules in the balloon cause a section of the vibrations to stop spreading out and instead come together (converge) at a point beyond the balloon (B). The increased vibrations at this point create a louder sound. Thus, the CO_2 gas in the balloon acts as a "lens" to focus sound!

The buzzer produces sound vibrations, and the carbon dioxide causes them to converge.

It's Not Black or White

PARENTAL SUPERVISION RECOMMENDED

Benham was a toymaker who lived in the 18th century. When he spun the black-and-white design you see below on a top, colors appeared! With this project, you can demonstrate this unusual phenomenon.

What You'll Need

- glue stick
- variable-speed electric drill
- sanding disk (for the drill)
- desk lamp with an incandescent bulb

Preparation

1. Copy the disk pattern shown here by photocopying it. Enlarge it as needed to make it the same size as the sanding disk.
2. Glue the disk pattern onto the sanding disk and attach it to the drill.

Presentation

1. Show your viewers Benham's disk. Then plug in the drill and the desk lamp and spin the disk under the bright light.
2. Ask your viewers what they see. The black-and-white disk is suddenly colorful! What colors do your viewers see, and what order are they in?

Why?

The retina in your eye has three types of receptors (areas sensitive to light). Each receptor is sensitive to a different color of light: one blue, one red, and one green. Not only do these receptors respond to different colors, but they also behave differently. For instance, when hit by light, the blue receptors take a longer time to send signals to your brain than the red or green receptors do. And once the blue or green receptors start sending signals, they take longer than the red receptors to stop sending signals.

When a steady white light hits your eyes, the slower receptors have time to catch up. After a short time, all receptors are responding equally, so you see the white (see The Shadow Knows!, page 4). But with Benham's disk, the white quickly flashes on and off as the disk spins. So, your slow receptors can't catch up with your fast ones. The result is that your color receptors never have a chance to respond equally, so instead of seeing white you see colors.

"X" Marks the Spot

PARENTAL SUPERVISION RECOMMENDED
Have you ever heard the saying "What goes up must come down"? Those words are talking about *gravity,* the force that keeps us from "falling" off the earth. Every object has a *center of gravity* (a point around which the pull of gravity is the same). This project shows you how you can find it.

What You'll Need

- string
- several large pieces of cardboard
- hole punch
- weight (a lead fishing kind will do)
- coat hanger
- two to three large, heavy books

Preparation

1. Cut the pieces of cardboard into various uneven and jagged shapes. *NOTE: You can also ask your viewers to cut the shapes. If they do so, have one shape cut already to give them ideas.*
2. With the hole punch (or the end of a pair of scissors), carefully punch two or three small holes around the edge of each shape (A).

3. Straighten out the hook on the hanger, then place the hanger under some books on a table so that the hook sticks out over the table's edge. The weight of the books will keep the hanger from falling (B).
4. Attach the weight to one end of the string, and tie a loop in the other end.

Presentation

1. Ask a person from the audience to hang a shape on the wire.
2. Then hang the looped end of the string on the wire in front of the shape. The weight will keep the string straight (B).
3. With a pencil, draw a line along the weighted string as shown.
4. Now hang the card by each of the other holes, each time drawing a line along the string. What do you notice? The lines all intersect at the same point—the center of gravity.
5. Show your viewers another shape and have them guess where the center of gravity might be. Then repeat steps 2 through 4 and see who's right.

Why?

For an object to balance on a single point, gravity must pull equally everywhere around this point. Since gravity pulls harder the more material there is, the balance point must have as much material (mass) on one side of it as on the other side of it. For example, on a long thin rod, each point along the rod has material to the left and right of it. What point has as much "rod material" to the right as to the left? The middle point, of course. But with your weird shapes, each point on the shape has material to the left and right and above and below it. To figure out just how much material there is left and right, and above and below, you need two pieces of information. The string with the hanging weight shows the direction of the earth's pull. When you hang the shape from the first hole, it will rotate until there are equal amounts of material to the left *and* right of the string. Hanging the shape from a second hole separates material equally in the up/down direction. Where the lines cross, there is the same amount of mass all around the intersection point (C). That's the center of gravity.

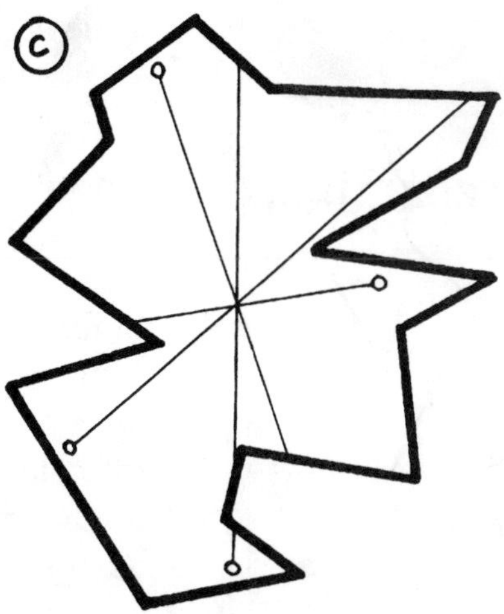

The center of gravity is where lines intersect.

Now think about an irregular cube shape. How would you figure out its center of gravity? You would need at least three lines to separate the material equally—an up/down line, a left/right one, and a front/back one.

How Sweet It Is

When you taste a food, you can tell if it has sugar (glucose) in it, right? Not necessarily! Just as scientists use telescopes to see further than their eyes can see, you can use chemical detectors to go beyond the limits of your ability to taste.

What You'll Need

- small paper cups
- stirring sticks
- watch with second hand
- garbage bag
- urine glucose test strips (available at drugstores)
- plastic spoons
- paper plates
- chalkboard and chalk
- poster-making materials
- the following food items: table sugar, steak sauce, honey, mashed potato, mashed beet, diet cola, a mashed green vegetable (such as zucchini), milk, mashed bread, fruit juice

Preparation

1. Create an interesting chart on a chalkboard to record the data you collect from the viewers of your project. Make a column for their names and columns for the different types of foods and for how sweet the viewers think each is.
2. On a poster, make a large version of the color code chart that comes on the side of the box of test strips.
3. Put a couple of teaspoons of each food into a separate paper cup. Rinse and dry the teaspoon each time you use it (A).

Presentation

1. Give each of three volunteers a plate containing a sample of your foods. Then have each volunteer taste the foods and rate them according to sweetness level. Write their answers down on your chalkboard.
2. Add a spoon of water to each of the paper cups you filled and stir.
3. Now hand out test strips to other viewers, and have them each dip the blue end of a strip into a cup of food. After 30 seconds, have them hold up their strips and show one another. Was glucose present in each of the foods? How correct were your volunteers' guesses?

Why?

In your project, when the glucose in various foods combines with the test strips, a chemical detector in the strips changes their color. The amount of glucose in a food determines the color that is created. You can use this fact to find out how much glucose is contained in different foods.

That's Poor Conduct

Ski jackets, sleeping bags, and even houses keep us warm because they contain a material called *insulation*. Do all materials insulate the same way? Tackle this project and find out.

What You'll Need

- 12 large plastic sandwich bags (the kind that "zip lock")
- equal portions of the following insulation materials: wool, flannel, human hair (available at a barber shop), cotton, and feathers from an old pillow
- ice chest with ice
- rubber bands
- watch with second hand
- heat lamp

Preparation

1. Start by making six "double bags." Turn one bag inside out, then slip it into a second bag so that the edges of the bags can lock together later (A). Repeat this with the remaining bags.
2. Fill five of the double bags with about 1" of insulation material, a different material per bag. Leave the sixth bag empty.
3. Lock the edges of each set of bags (B). For your presentation, have the bags, the filled ice chest, and the heat lamp ready.

A
bag turned inside out
bag right side out

Presentation

1. Ask a volunteer to choose two insulation "gloves" to wear. Put rubber bands around the person's wrists to hold the gloves closed.
2. Check your watch for the time, then ask the volunteer to put both hands first in the ice chest, then near the heat lamp. How long does it take for each hand to feel the cold? To feel the warmth?
3. Repeat steps 1 and 2 with two other volunteers, so that all five insulation materials, as well as the empty double bag, are tested. Which insulation works the best?

B
After adding insulation material, lock the inner bag to the outer one.

Why?

The transfer of heat is a much slower process in insulators than in metals (see River of Heat, page 32). Also, many of the materials you used, like wool and cotton, trap pockets of air between their fibers. Heat travels slowly in air because, unlike in solids, the molecules are much farther apart. The empty double bag of plastic insulated well not only because plastic is a good insulator, but also because there is air between the bags.

'Round the Bend

PARENTAL SUPERVISION RECOMMENDED

If you stand a pencil in a glass of water, then look at it from the side, the pencil looks strange. The part in the water seems to have moved *next* to the part out of water! Actually, it's not the pencil that's bent, but the light reflecting off of it. We call bending light *refraction*. In this project, you'll see refraction at work.

What You'll Need

- talcum powder and puff
- mirror
- cardboard, 1' x 1'
- sturdy cardboard box
- two pieces of wood, 2" x 2" x 6"
- deep glass dish
- milk
- flashlight
- tape
- eyedropper

Preparation

1. About one-third of the way down the piece of cardboard, cut a slit 6" long and $\frac{1}{2}$" high (A).
2. Prop up the cardboard on end between the pieces of wood.
3. Put some books behind one flap on top of the cardboard box to create an angled surface as shown (B).
4. Tape the flashlight to the top of the box with the light pointed down. Make sure the light shines through the cardboard slit.
5. Put the mirror in the bottom of the dish, then fill the dish with water.

A

slit in cardboard

block of wood

B

The light beam refracts, reflects, then refracts again.

tape holds flashlight in place

mirror

cardboard stands between blocks of wood

Presentation

1. Turn on the flashlight, then put a couple of drops of milk in the water. Ask your viewers if they can see where the light beam is going.
2. Now pat some powder in the air along the light beam's path (B). This will make the beam show up, and your viewers will be able to see it bending.

Why?

The best way to understand why the light beam bends as it enters and exits the water is to perform this simple experiment. Get a rolling pin (or something similar) and set it rolling across a table. But push one end of it faster than the other. Notice how the rolling pin turns. Only when both ends are pushed at the same speed and with the same force does the rolling pin travel straight.

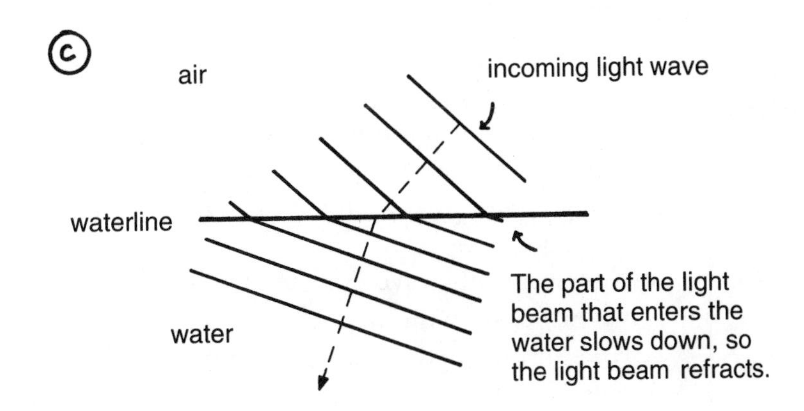

The same thing is happening to your light beam. Light travels slower in water than it does in air. When the beam hits the water at an angle, the side entering the water slows down. And just like the rolling pin, the light beam "bends," or refracts (C). Once the beam is completely inside the water, all parts of it travel at the same speed. Then, as the light exits the water (after reflecting off the mirror), it immediately speeds up again now that it's back in the air. That speeding up refracts the light beam once more.

Disappearing Act

PARENTAL SUPERVISION RECOMMENDED
With this project, you can use the powers of refraction in a magical way. Your friends won't believe their eyes!

What You'll Need

- Wesson™ oil
- large beaker
- hammer
- tongs
- two small beakers, each small enough to fit inside the large beaker
- handkerchief
- glass jar
- towel

Preparation

1. Place one of the small beakers into the large one.
2. Now hide the small beaker by filling the large one to the top with oil.

Presentation

1. Show your viewers the other small beaker. Tell them you will break it, then magically put it back together again.
2. Wrap the beaker in a towel, then smash it with the hammer.
3. Using the tongs, drop the broken pieces into the large beaker (not the small one).
4. Now magically wave the handkerchief over the beaker and pull out the whole beaker with the tongs, leaving the broken pieces behind. Can your viewers figure out how you did this "magical" feat? (If anyone doubts you, pour the oil out of the large beaker into the jar!)

Why?

We see opaque (nontransparent) objects by the way they reflect light. But light simply passes through transparent objects, such as glass or plastic, so why can we see them?

When light passes from one material to another (say, from a table to a glass door), its speed often changes, so the light bends. The bending of light also distorts (or changes) the appearance of objects behind a transparent one. So, we can tell if glass is there by the distortion of objects behind it.

In your project, the beakers and the oil have something in common: light travels at the same speed through both of them. That means that once light enters the oil, it travels straight through the beaker without bending, and that makes the broken glass totally invisible!

Bubbles, Bubbles Everywhere

PARENTAL SUPERVISION RECOMMENDED

How are soap bubbles similar to the cells in our bodies? Do this project and find out.

What You'll Need

- Plexiglas™ box, about 1' wide, 1' deep, and 3' tall (an emptied aquarium will work well, too)
- dry ice
- plastic insulated gloves
- 2/3 cup of dishwashing liquid
- 1 tablespoon of glycerine (available at drugstores)
- 1 gallon of water
- large mixing bowl (or bucket)
- wire
- dishcloth

Preparation

1. At a plastics or hardware store, have some sheets of Plexiglas cut and glued together into a tall box.
2. Make a bubble-blowing solution by mixing the dishwashing liquid, glycerine, and water in the bowl. Let the mixture sit one day.
3. With the wire, make a 2-inch bubble blower.
4. Just before your demonstration, put the box on the floor next to a table, place the dishcloth in the bottom of the box, and put the dry ice on top of the cloth. *WARNING! Dry ice is extremely cold! Wear plastic insulated gloves while handling it. Do NOT touch it.*

(A)

Presentation

Ask someone from the audience to dip the bubble blower into the mixture, then blow bubbles into the box so that they sink to the bottom (A). What happens then? Do they burst? Do they get bigger?

Why?

Bubbles have what's called a semipermeable membrane. That means that their thin outer surfaces allow certain molecules to enter. One of these molecules is carbon dioxide (CO_2). As the bubbles sink into the CO_2 gas created by the dry ice, CO_2 molecules gradually diffuse into (enter) the bubbles (B). Air molecules, on the other hand, are not able to diffuse into or out of the bubbles. With no air leaking out of the bubbles, and with CO_2 leaking in, the bubbles grow in size.

(B)

The cells in your body also have a thin, permeable membrane surrounding them that allows only certain molecules (such as oxygen) in or out. So at least one aspect of cells can be understood by blowing bubbles!

Impossible Arch

Sometimes by using very simple and basic ideas, you can create astonishing effects. By understanding something about balance, you can construct an arch that seems to defy gravity.

What You'll Need

- 10 to 15 books (thin, hardbound books work best) or compact disk cases

Presentation

1. Stack the books evenly on top of each other.
2. Slide the top book (in the direction of its longest side) so it hangs over the book below it as far as possible without falling (A).
3. Then slide the second book (and the first book along with it) out over the third as far as possible. Continue sliding the books outward until you've reached the bottom.
4. You've finished your impossible arch. Look closely. What do you or your viewers notice? Why doesn't it fall over?
5. Measure the distance that you slid each book. As you move down the stack, you'll find that you slid each book half as far as the one above it.

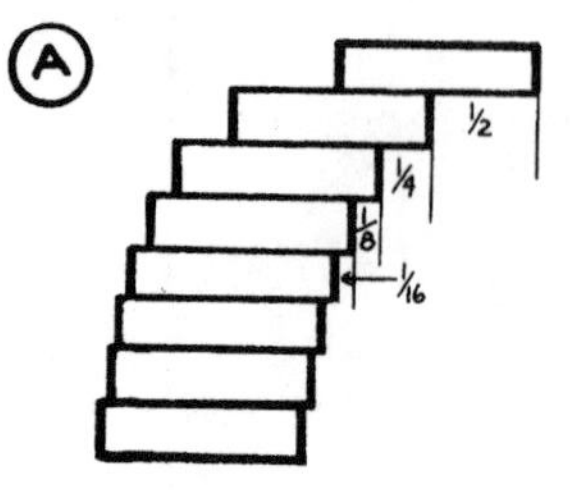

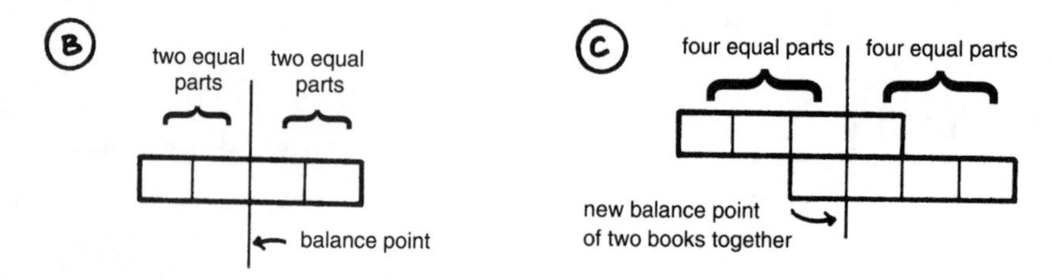

Why?

To balance an object, you must find the place that has an equal amount of mass on either side of it (see "X" Marks the Spot, page 44). Look at diagram (B) above. The balance point of the top book is in the center of the book, so it balances when its center is just on the edge of the second book. Now the first and second book together form a new object with its own balance point (C). The balance point of the two books together is in the middle of the region where the first and second books overlap (there's an equal amount of "parts" on either side of this point). Then the first, second, and third books form a new object with its own balance point, and so forth. By continuing in this way, you can see how the entire stack can be balanced.

Left Brain, Right Brain

PARENTAL SUPERVISION RECOMMENDED

Can you draw what you see in a mirror? With this project, your viewers can test their artistic talent with the help—or hindrance!—of a mirror.

What You'll Need

- flat (plane) mirror
- rearview (convex) mirror
- shaving (concave) mirror
- cardboard box
- pencil and paper
- double-sided tape
- sheet of cardboard, larger than one side of the box

Preparation

(A)

1. With the pencil and paper, draw three or more copies of the irregular shape,(each shape should measure about 6" from top to bottom). Draw a dark outer line and a thin inner line (A). *NOTE: Any shape can be used, as long as it has outer and inner lines.*
2. Cut out the top flaps and one side of the cardboard box.
3. With the tape, place the plane mirror against the far wall of the box, then lay the cardboard sheet over the box (B).
4. Put one copy of the irregular shape on the floor of the box. Make sure you can see the shape reflected in the mirror.

Presentation

1. Ask one of your viewers to look at the shape's reflection in the mirror. Then hand him the pencil and ask him to place the point between the inner and outer lines of the real shape, then to draw the shape between these lines without going outside the lines and without taking his eyes off the mirror. He'll probably have a really tough time doing it!
2. Now repeat step 1, but replace the plane mirror with the convex mirror, then with the concave mirror. Does anything different happen?

Why?

As you move the pencil away from you and toward the mirror, the image of the pencil *looks* closer to you. As a result, two different messages are reaching your brain as you try to draw—the signal from your hand saying the pencil is moving away from you, and the visual signal telling you the pencil is coming closer. The two conflicting signals confuse you and make it difficult to draw!

Paper Caper

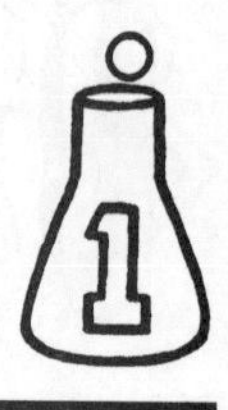

How many times do you think you can fold a piece of construction paper in half? Can you fold it up to ten times? With this project, you can bet your friends it's impossible!

What You'll Need

- a large sheet of colorful construction paper, about 28" x 22"

Presentation

1. Ask a viewer to try to fold the construction paper in half lengthwise, crosswise, or even diagonally.
2. Tell the viewer to keep folding the paper in half as many times as possible. He will not be able to fold the paper more than nine times, if that many!

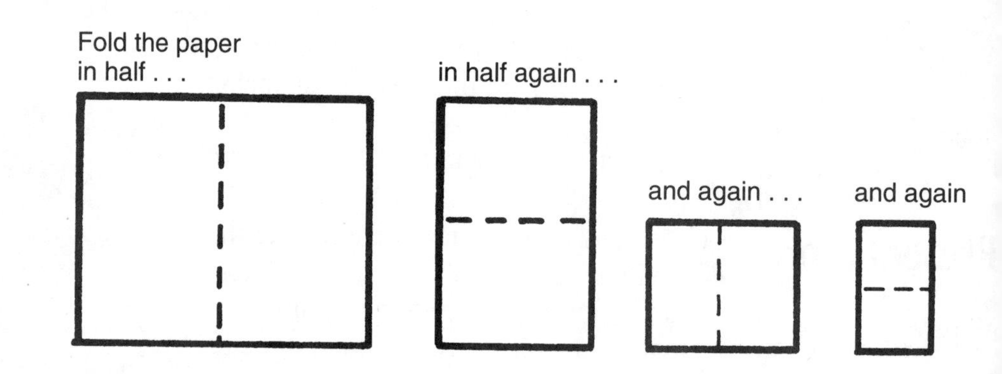

Why?

When you fold a piece of paper, the number of layers increases *geometrically*. That means the number of layers doubles with each fold. After the first fold, there are 2 layers of paper. After the second fold, there are 4 layers of paper, and after the third, 8 layers. By the sixth fold, there are 64 layers. The seventh fold produces 128 layers! You might as well be trying to fold a book in half!

Blend and Disappear

PARENTAL SUPERVISION RECOMMENDED

The last time you walked through the woods enjoying the scenery, were you startled to see a *plant* get up and walk away? You didn't realize it before, but you were looking at a camouflaged animal! Why can animals (and some plants, too, by the way) fool us like that? "Disappear" into this project and find out.

What You'll Need

- two pieces of black or dark blue construction paper, 3' x 3'
- two sheets of stiff cardboard, 3' x 3'
- glue
- liquid correction fluid or white model paint
- clear plastic sheet, 3' x 3'
- easel

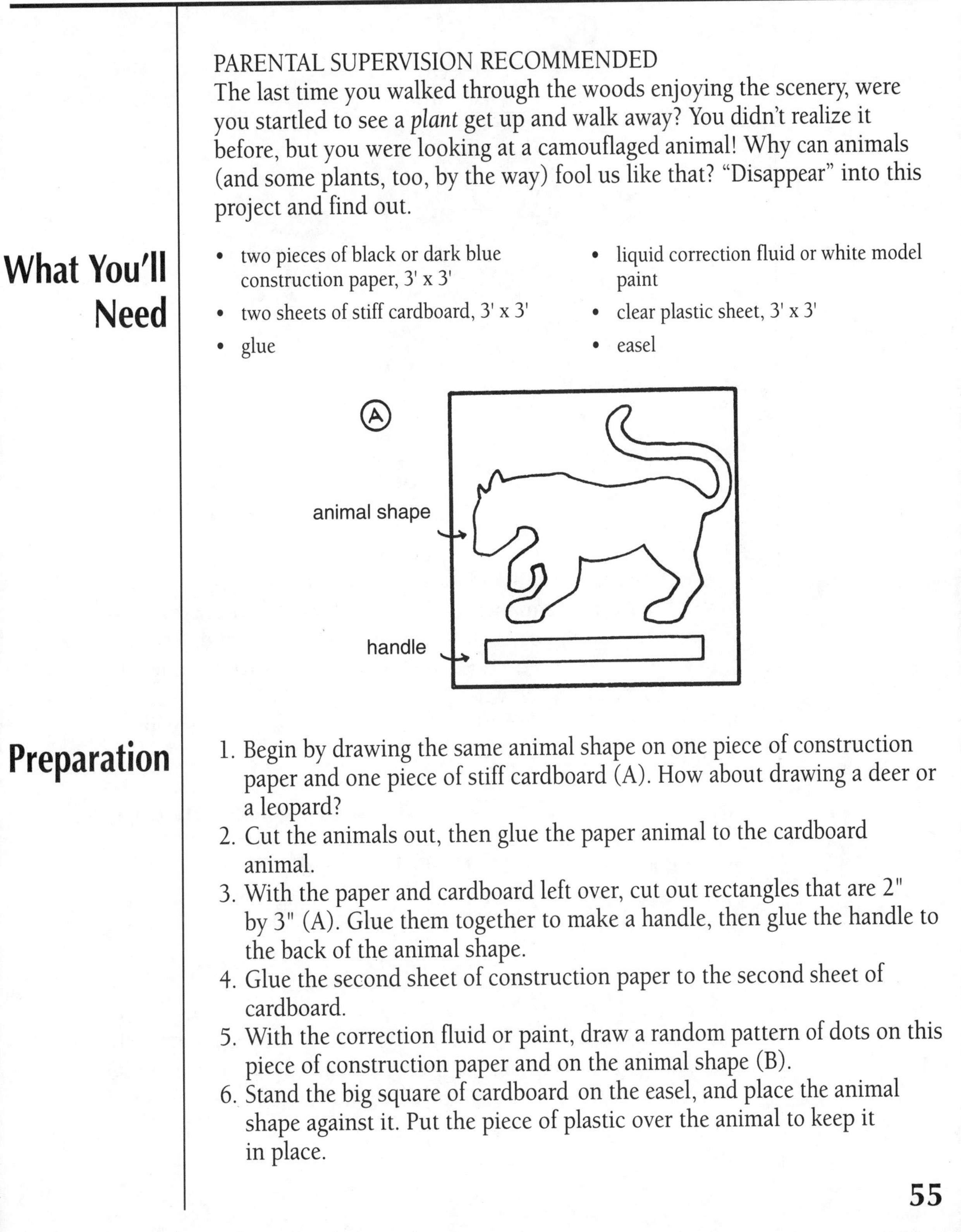

Preparation

1. Begin by drawing the same animal shape on one piece of construction paper and one piece of stiff cardboard (A). How about drawing a deer or a leopard?
2. Cut the animals out, then glue the paper animal to the cardboard animal.
3. With the paper and cardboard left over, cut out rectangles that are 2" by 3" (A). Glue them together to make a handle, then glue the handle to the back of the animal shape.
4. Glue the second sheet of construction paper to the second sheet of cardboard.
5. With the correction fluid or paint, draw a random pattern of dots on this piece of construction paper and on the animal shape (B).
6. Stand the big square of cardboard on the easel, and place the animal shape against it. Put the piece of plastic over the animal to keep it in place.

B

Presentation

1. Ask the audience to stand about 6 feet away from your display.
2. Ask them if they can identify any kind of shape on the cardboard.
3. After they have guessed, lift the plastic a little bit, and use the handle to move the animal around. Say, "There's an animal lurking around here. Can anyone tell what it is?"
4. Move the animal away from its "surroundings" to show your viewers.

Why?

Since life began developing on earth roughly 3 billion years ago, the types of creatures that survived were those best suited to their environments. Animals that blended in with their surroundings could hide from predators or sneak up on prey, making their survival likely. As a result, many animals now exist that have colors and patterns on their bodies that match the colors and patterns of the plants, rocks, and soil where they live. When motionless, these animals are hard to see. The reason you can see them clearly when they move is because many animals, including humans, have developed specialized brain cells that can detect changes in the light patterns received by the eye. Motion changes the patterns you see, so you can see an animal in motion.

Spinning Wheel

In this project, you can take your friends for a spin and in the process learn something about reaction forces!

What You'll Need

- bicycle wheel
- rotating stool or chair
- two plastic file handles (available at hardware stores)

Preparation

1. Put the plastic handles on the axle of the bicycle wheel. Make sure the wheel spins easily.
2. Have the chair and wheel ready for your presentation.

Presentation

1. Start by holding the wheel by the handles. Ask someone in the audience to spin the wheel as fast as possible. You have just created a gyroscope.
2. Then quickly sit on the stool, lift your feet off the floor, and tilt the wheel. The stool will start to turn (A).
3. Now tilt the wheel in the other direction. What happens?
4. For fun, invite volunteers to sit on the stool and hold the spinning bicycle wheel.

Why?

Whenever you try to change the way something is moving, you feel a reaction force on you. That's because for every action there is an equal and opposite *re*action. For example, when you hit a moving baseball, you feel the crack of the bat push against you. Now think about each part of the bicycle wheel in your project. When spinning, each part is moving in a certain circular path. When you tilt the file handles, you are forcing all the parts of the wheel to travel along a *different* circular path (B). In other words, you are changing the way the wheel is moving. So, you feel a reaction force that spins you in your chair!

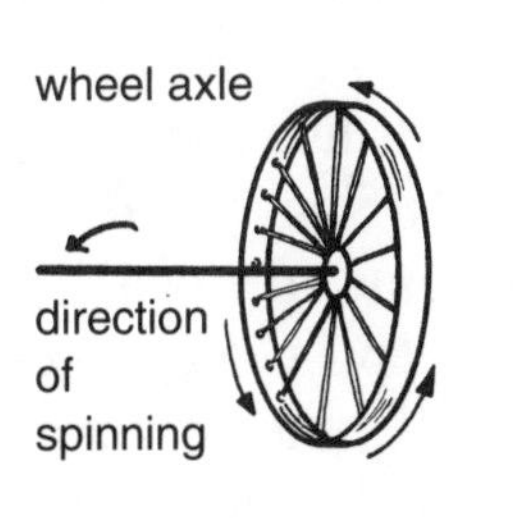

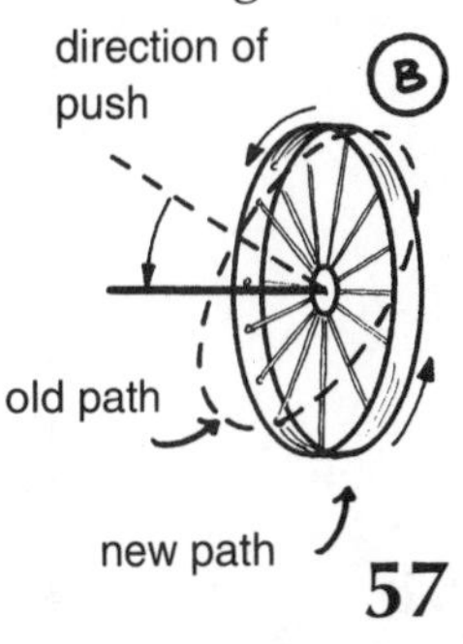

Static Magic

Have you ever dragged your feet across carpeting, then touched something made of metal and received a tiny electric shock? If you have, you know the effects of *static electricity.* With this project, you can demonstrate how static electricity works by creating your own dancing "peanuts"!

What You'll Need

- sheet of Plexiglas™, 1' to 2' square
- four blocks of wood
- glue
- paper bag
- Styrofoam™ "packing peanuts" (the bits of plastic used as cushioning in packages)
- wool cloth (or sweater)

Preparation

1. Glue the four blocks to the corners of the Plexiglas sheet. Let dry.
2. Put the packing peanuts in the paper bag.

Presentation

1. Ask your viewers to watch while you rub the Plexiglas with the wool (A). Tell them you are charging it with static electricity.
2. Now pour the packing peanuts onto the Plexiglas (B). Within just a few seconds, the Styrofoam peanuts will start jumping.
3. Rub the Plexiglas again and ask for a volunteer with long hair to hold her head near the sheet. Her hair will stick to the Plexiglas!

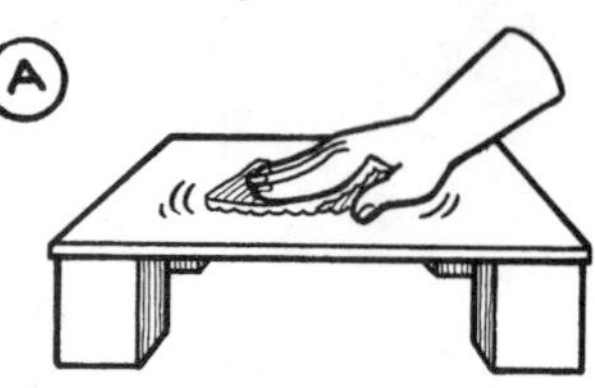

Why?

The atoms that make up all the different things around you are built out of microscopic particles that have a property called *charge*. There are two types of charged particles: positive and negative. While like charges attract (pull towards) each other, opposite charges repel (push away from) each other. Most objects around you contain equal numbers of both positive and negative charges, so their effects cancel out. In your project, though, when you rub the wool on the Plexiglas you actually strip negative charges off the wool and put them onto the plastic. As a result, the plastic becomes *negatively charged.* When the Styrofoam peanuts are then dropped onto the plastic, some of these excess negative charges leak onto the peanuts, making them negatively charged, too. Since the plastic and peanuts now have an excess of the same charge, they repel each other—and the peanuts jump away!

The same negative charges on the plastic attract the positive charges in your volunteer's hair, pulling it toward the plastic.

Magnetic Levitation

PARENTAL SUPERVISION RECOMMENDED

Two famous scientists, Michael Faraday and James Clerk Maxwell, discovered that electricity and magnetism were not completely different forces, but actually closely related effects. In this simple project, you can explore this amazing connection.

What You'll Need

- two aluminum sheets, about 8" x 11" and at least $\frac{1}{4}$" thick (available at metal shops or hardware stores)
- a strong, donut-shaped magnet, about 2" in diameter and $\frac{1}{4}$" thick (a neodymium magnet works well)
- utility knife
- disk of Styrofoam™, cut to the exact diameter and thickness of the magnet
- two Styrofoam strips, 1" x 10" and at least $\frac{1}{4}$" thicker than the magnet
- thin white posterboard
- rubber cement

Preparation

1. Cut four circles out of the posterboard that are the same diameter as the magnet. Glue the circles to both sides of the magnet and the Styrofoam disk. Then cut and glue a strip of posterboard to cover the sides of the magnet and Styrofoam (A). These two disks should now look the same.
2. Write "ANTIGRAVITY" on the disk with the magnet inside. Write "GRAVITY" on the other disk.
3. Now glue the two Styrofoam strips to the edges of one aluminum sheet (B). Glue the second aluminum sheet on top and let dry.

(A)

ANTIGRAVITY

Cover magnet (and Styrofoam disk, too) with posterboard.

(B)

aluminum sheet

Styrofoam strip

Presentation

1. Have your viewers feel the weight of each disk, then tell them that even though the disk marked ANTIGRAVITY is heavier, it will fall more slowly than the other one, as if it is fighting gravity.
2. While holding the aluminum sheets vertically about a foot above a table, have a viewer drop both disks between the aluminum sheets at the same time (C). The GRAVITY disk will fall through quickly, while the ANTI-GRAVITY disk will fall very slowly. Ask your viewers to guess why!

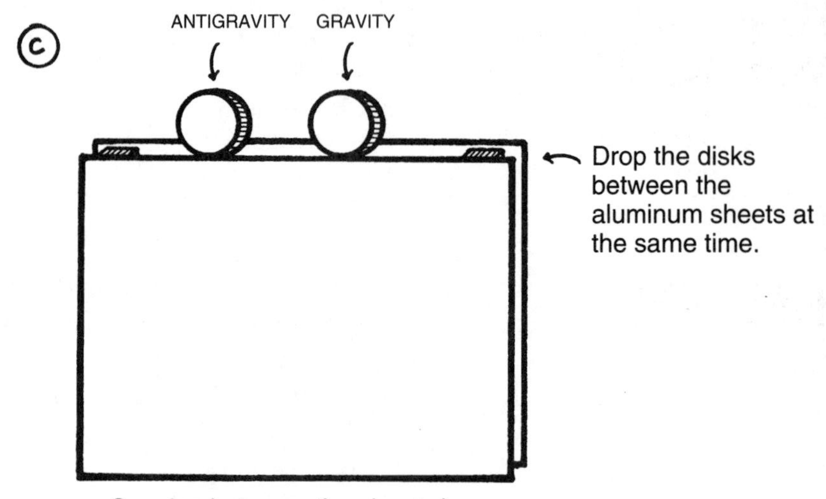

Why?

Remember that a charged particle creates an invisible electric force field around it (see Static Magic, page 58). That force field attracts or repels other charged particles around it. When you move a charged particle, the field moves with it and creates a magnetic field. The reverse happens, too! When you move a magnet, its magnetic field also moves and in so doing creates an electric field (just like the one created by a charged particle).

In your project, as the magnetic disk falls, it creates an electric field. The field causes charged particles in the aluminum to move. These moving charges in turn create magnetic fields that repel the falling magnet, slowing its fall.

Electricity and magnetism are not completely different effects. They are closely related. One can create the other, so are they different at all?

Half-Life Dating

When scientists discover fossilized bones (like those of a saber-toothed cat), how do they know how old the remains are? One way is by *half-life dating*. Some radioactive elements, such as carbon 14, decay over a period of time. A half-life is the time it takes half the atoms in a chunk of radioactive material to decay into new atoms. With this project, you can demonstrate just how half-life dating is an important tool for scientists.

What You'll Need

- 100 poker chips (or coins), all the same color
- large sheet of white construction paper (or posterboard)
- empty box
- sheet of acetate, the same size as the construction paper
- paint and paintbrush
- two marking pens, one black and one a color that will show up against white

Preparation

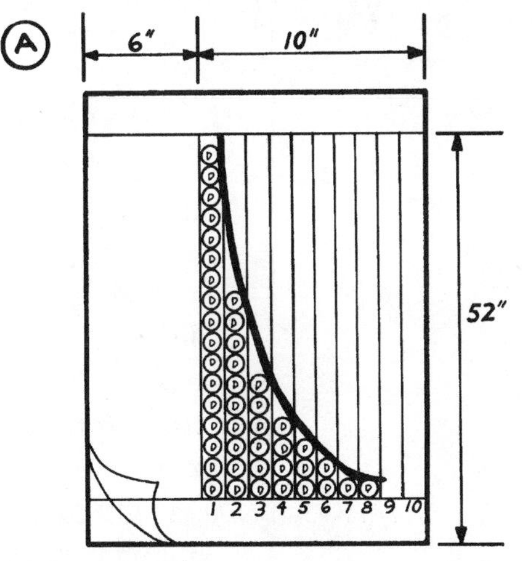

1. Start by painting "D" (for "Decayed") on one side of each of the poker chips and "ND" (for "Not Decayed") on the other side.
2. When the chips are dry, put them into the box.
3. Using the large sheet of construction paper and the colored pen, make a chart similar to the one shown (A).
4. Before your presentation, determine the *result of a set of trials* in the following way. Toss the chips onto a table (B). Then place the chips that come out "D" in the first column on your chart. Put the remaining chips back into the box, toss them again, then place the "D" chips in the second column. Repeat the tossing until all the chips are gone or you have run ten trials. *NOTE: If no chips turn up "D" for a particular trial, leave the column for that trial empty.*
5. The result of this set of trials is the curve the "D" chips make. Lay the acetate over the chart, then draw this curve above the chips using the black pen (A).
6. Set the sheet of acetate aside and put all the chips back in the box.

(B)

Presentation

1. Tell your audience that each chip in the box has a "D" or "ND" on it and what these abbreviations mean. Then describe what radioactive decay is and that the chips represent the number of radioactive atoms in a fossil. Tell them that each of these "atoms" may or may not decay during your trials, but that you know what the results of the trials will be before you even perform them.
2. Now repeat the set of trials you performed in step 4 above. Explain to your audience that since the chips have only two sides, there is a 50-50 chance they will be "D" or "ND." When you're done with the tossings, place the acetate down on the chips. The curve you drew earlier will almost perfectly fit the new set of trials!

Why?

When you flip a penny, although you have no idea whether it will come up heads or tails, over many flips you can predict the penny will land heads up or tails up about the same number of times. There are two reasons why you can predict something about a large number of flips but nothing about a single flip. The first is that you know there are only two possible outcomes (heads, tails). The second is that which outcome you get is completely random—neither result is more likely. If many more heads showed up in a few tossings of a coin, you'd suspect that something was causing heads to come up more often. If the coin were magnetic, or had heads on both sides (in other words, the coin is fake), then the possible outcome is no longer purely random. As a result, you couldn't predict a long-term result anymore. But when the process is completely random (as it is with pennies, and decaying atoms, too), then you'll find about an equal number of both possible outcomes. This idea allows you to predict what will happen to a large number of flipped pennies, decaying atoms, and many more things.

The curve you drew on the acetate was a prediction of what would happen. That same curve could predict what would happen in any random system that has only two possible outcomes. Because scientists know how fast atoms decay, they can tell how old something is by counting how many atoms have decayed. In the same way, you can tell from your curve how many trials have passed by counting the number of "D" chips in *any* given column. Did you notice that with every dumping of the box of chips, about half of them comes up "D"? You can see why, when dealing with decaying atoms, the process is called half-life dating.

Wayward Compasses

PARENTAL SUPERVISION RECOMMENDED
If you've ever used a compass before, you probably noticed that the needle inside always points north. Is it possible to make the needle point in another direction? This project will show that it is.

What You'll Need

- 6- or 12-volt battery
- large cardboard box
- sheet of cardboard, about 8" x 8"
- two rulers
- two wedge-shaped pieces of wood
- glue
- wire coat hanger
- six small compasses about 1" in diameter
- two electric lead wires with alligator clips at one end
- wire cutter

Preparation

1. Start by cutting off the top and one side of the box. Then cut two holes in the left side of the box as shown (A).

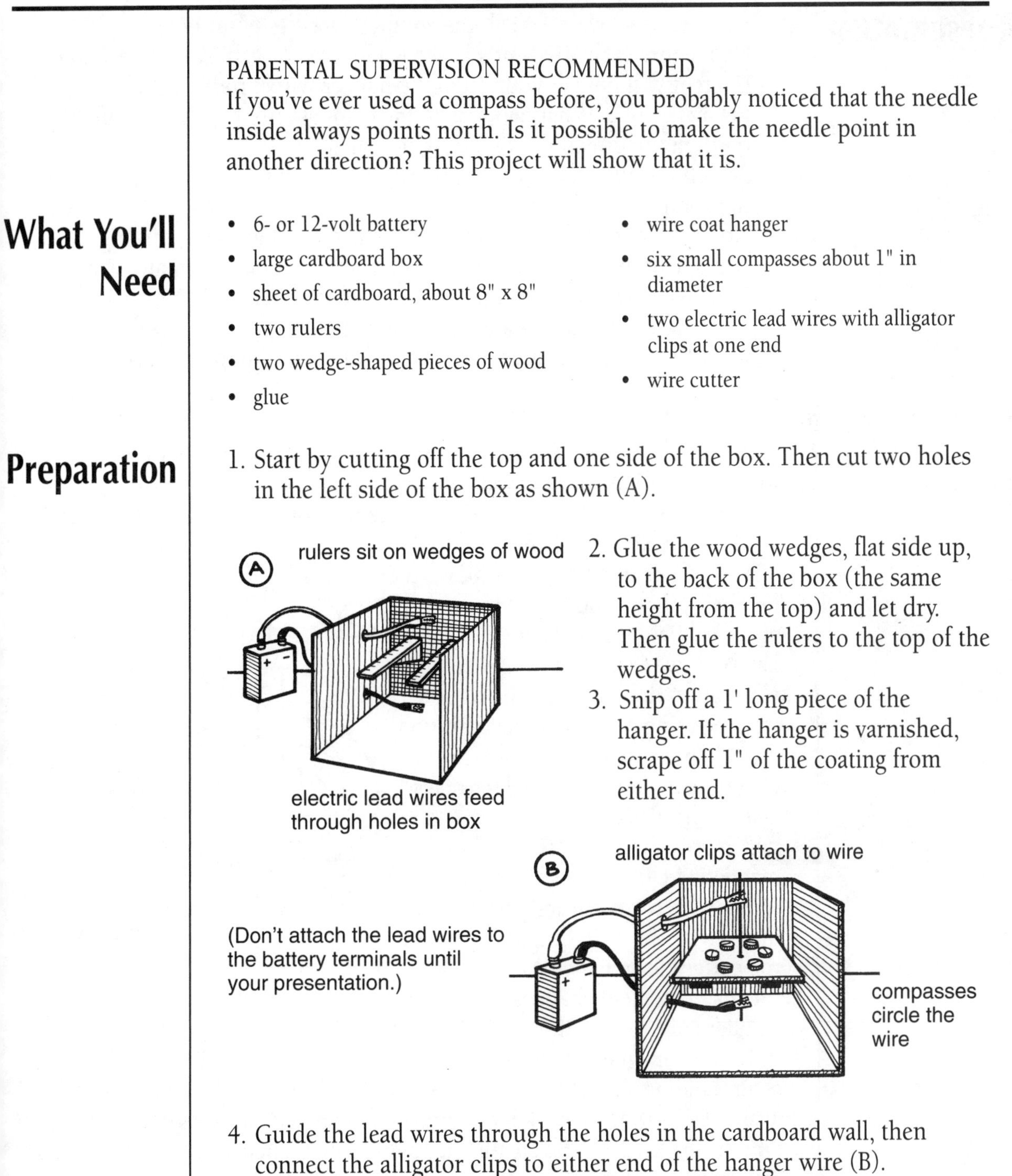

2. Glue the wood wedges, flat side up, to the back of the box (the same height from the top) and let dry. Then glue the rulers to the top of the wedges.
3. Snip off a 1' long piece of the hanger. If the hanger is varnished, scrape off 1" of the coating from either end.
4. Guide the lead wires through the holes in the cardboard wall, then connect the alligator clips to either end of the hanger wire (B).
5. Circle the compasses around the wire on the cardboard sheet.

Presentation

1. Show your viewers that all the compass needles point north.
2. Next connect the free ends of the lead wires to each battery terminal. Current will now flow through the hanger wire, producing a magnetic field. What happens to the compass needles? They will all point in the direction of the magnetic field!
3. Switch the alligator clips on the battery terminals. What happens to the compass needles now?
4. Now take one of the compasses (one that is not already pointing in a northerly direction), and, while your viewers accompany you, walk away from your setup. Have everyone watch what happens to the compass needle. At a certain point, the magnetic field will be too weak to affect the compass and the needle will flip to the north.

Why?

Why did the compasses stop pointing north when the wires were hooked to the battery? What changed? An electric current was now flowing through the wires. Electric current is the flow of charged particles. These particles create an invisible force field around them (see Static Magic, page 58). When charged particles move, they drag this force field with them, and the changing electrical field is a magnetic field. The battery in your project causes charged particles to flow through the hanger wire, creating a magnetic field around it (C). The compass needles also contain circulating charged particles. These particles create tiny magnets, which are attracted to the wire's magnetic field, causing the needles to rotate. When you reverse the current in the wire (by switching the clips on the battery terminals), the direction of the magnetic field changes, so the needles change direction, too.

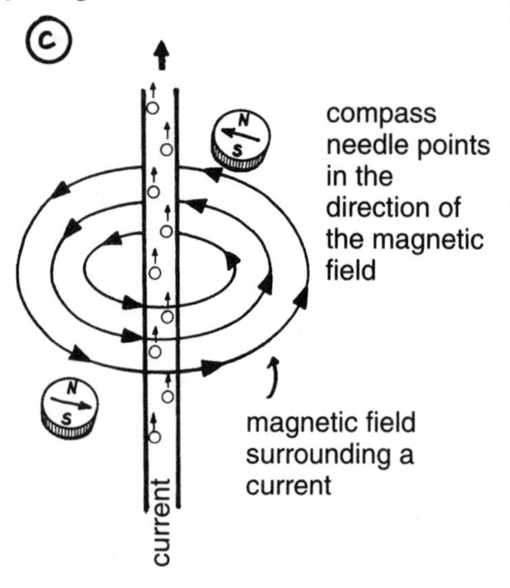

When a magnetic field is very weak, as when you walk away from your setup holding the compass, why does the compass needle point north again? Deep in the earth's molten interior, charged particles circulate around the earth's core, creating a huge magnetic field that spans the globe. This magnetic field causes all compasses to point north.

40

Mirror, Mirror

When we look at a mirror, we see a *reversed* image of ourselves (our left is on the right and vice-versa). Why is that? Why isn't the image also upside-down? Can it be? With this project, you can demonstrate the answer.

What You'll Need

- flat (plane) mirror, about 8" x 12" or larger, with a stand
- block of wood, 4" x 4" x 1'
- three large Styrofoam™ balls
- white and red paints
- large and small paintbrushes

Preparation

1. Begin by painting the wood red, using the large brush. Let dry.
2. Next, using the small brush and the white paint, paint your name down one side of the wood and the words "BLUE BOX" down the other side.

Presentation

1. Place the plane mirror on a table and stand the wood block on it so it reflects your name (A).
2. Then prop the plane mirror up and put the block upright next to it (B).
3. Lay the wood down in front of the mirror so that "BLUE BOX" faces up (C). Notice that "BOX" is readable. Ask your viewers why.

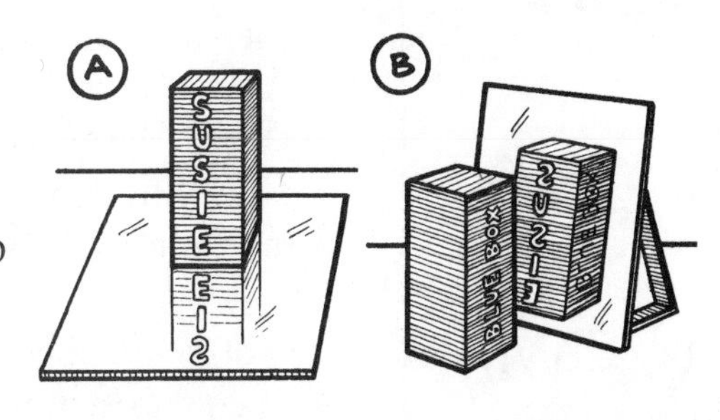

Why?

To try to understand what's going on, try a simple observation. Stand with your hands raised up in front of you, palms facing out. Have a friend stand facing you, doing the same. Your hands will look like mirror images of each other. We see objects when they reflect light into our eyes. Notice that, in traveling along a line toward your eye, light from your friend's *left* hand reaches your eye from your right. Now stand in front of a mirror the same way and look at the image of your right hand. For you to see your right hand, light reflecting off of it must travel to the mirror, then bounce off the mirror and back to your eyes. Notice that this light from the mirror (from your right hand) travels along the same path as light from your friend's left hand—namely, from your right side! So, light bouncing back toward you from your right side is like light coming straight at you from your friend's left side. Well, your brain can't tell when light has been bounced off a mirror and when it hasn't. So, it sees light reflected from the right as coming from the left. This is why mirror images are reversed left to right.

Flying High

A jumbo jet rolls down the runway, picks up speed, and within seconds has lifted off the ground and is flying through the air. How did the airplane get off the ground? In this project, you can learn how.

What You'll Need

- hair dryer (or vacuum cleaner with reversible hose)
- wall area
- Ping-Pong ball (or beach ball)

Preparation

Set up your hair dryer and Ping-Pong ball near a wall.

Presentation

1. Turn on the hair dryer, aim it toward the ceiling, and balance the ball in the airstream above it.
2. Ask a volunteer to pull the ball partially out of the airstream, then slowly let go. The ball will get sucked back into the airstream!
3. Now move close to a wall and balance the ball in the airstream. Notice anything happen?

Why?

Do you remember that for every action there is an equal reaction (see Spinning Wheel, page 57)? Air molecules forced upward by the hair dryer eventually collide with the ball and are deflected (pushed) downward. The ball, in deflecting air downward, experiences a reaction force upward, causing it to float. (In airplanes, air colliding with the tilted wings is deflected downward, causing the plane to experience a reaction force and rise.) When you pull the ball halfway out of the airstream, air molecules on the stream side collide with the ball and deflect away. As they do so, they sweep away other air molecules nearby. The ball now has less air pushing on it from the stream side than from the outside (where there are constant collisions). As a result, the ball gets pushed back into the airstream.

airstream curves around ball

Why does the ball rise higher near a wall? Normally, as the airstream moves up, it spreads out and becomes weaker. By bringing the airstream close to a wall, the air doesn't have room to spread out, so it stays more concentrated and the ball rises higher.

Only Your Cabbage Knows

This project experiments with chemistry. By adding different products to an "indicator" made of cabbage juice, you can turn things green!

What You'll Need

- 2 or 3 cans of cooked red cabbage
- the following products to test: baking soda, vinegar, ammonia, club soda, cream of tartar, carrot juice, apple juice, tabasco sauce, beet juice, ground aspirin
- 6 or 7 glasses
- pitcher
- plastic tablecloth
- wooden stirrers
- eyedropper

Preparation

Just before your presentation, cover a table with the tablecloth. Pour about 1/4 cup of juice from the cans of cabbage into each of the glasses. Display the cans or bottles that your test products come in next to the glasses.

Presentation

1. From your research, explain to your viewers the difference between acids and bases. To help them understand, tell them that milk is a base and orange juice is an acid. The cabbage juice is an indicator.
2. Ask several volunteers to look at the test products and guess which ones are acids and which are bases. Ask them to write down their guesses.
3. Add a pinch of the baking soda into the first glass of cabbage indicator and stir. The color will change to green (if it doesn't, add more baking soda). The green shows that the mixture is a base.
4. Now add about half a teaspoon of vinegar to the second glass. The mixture will remain red, showing that it is an acid.
5. Repeat with the remaining products and glasses. Which are acids and which are bases? For fun, take the indicator that the baking soda was added to and add some vinegar. What happens to the mixture?

Why?

A compound consists of two or more kinds of atoms that combine chemically. The simplest atom is hydrogen. Two groups of compounds that are very easy to identify are *acids* and *bases*. Acids readily *give away* hydrogen atoms to other compounds, while bases readily *grab* hydrogen atoms. When you pour an acid into a base (or vice-versa), they neutralize each other. Why? Because the base grabs the hydrogen atoms that the acid wants to give away anyway!

Spare Me the Details

PARENTAL SUPERVISION RECOMMENDED

Most of what happens around us is the result of many complicated interactions that we can't keep track of. Fortunately, we can often ignore most of the details and look at the "big picture." For example, while we can't possibly know the detailed motion of a single water molecule in a river, we can still predict something about the motion of the river as a whole. In this project, you'll create a device that is a good model for many systems in nature.

What You'll Need

- two rigid Plexiglas™ sheets, 17" x 16" and 1/8" thick
- two pieces of Plexiglas, 17" x 1/2" and 1/8" thick
- one piece of Plexiglas, 16" x 1/2" and 1/8" thick
- 160 plastic pegs, 1/2" long and 1/4" in diameter
- 14 rectangular Plexiglas pieces, 1/2" x 8" and 1/8" thick
- plastic glue
- marker that can be erased from plastic
- 200 small marbles (or Pachinco balls), 1/4" in diameter
- funnel large enough for the marbles to easily fall through

Preparation

1. Start by making a pegboard, 8 rows deep and 20 rows across (A). From left to right, the pegs should be spaced 3/4" apart (from the center of one peg to the next). From top to bottom, the rows should be 1" apart (again from the center of one peg to the next)(B). Shift every other row to the right 3/8", so that each peg is *between* the two above it. Glue the rods in place and let dry.
2. Glue down the 14 rectangular Plexiglas "slot" pieces on their edges as shown, 1" apart.
3. Glue on the top and side Plexiglas pieces to complete the box as shown. Allow the entire assembly to dry for several hours.

Presentation

1. Stand your box vertically, with the open end up. While a volunteer holds the funnel, pour the 200 marbles through the funnel into the center of the box (B). Now the fun begins!

(A)

clear plastic box

plastic pegs

(illustration not to scale)

Plastic rectangles glued on their sides form bins for marbles to fill.

2. The marbles will collide with the pegs on the way down and get knocked in all different directions. Eventually, all the marbles will fall into one of the slots below.
3. With the marker, draw a curve on the plastic along the tops of the stacked marbles. Notice how the curve is shaped like a bell—high in the middle and dropping off toward the edges.
4. Pour the balls out and repeat step 1. When the marbles are done falling, they should form the same shape as before.

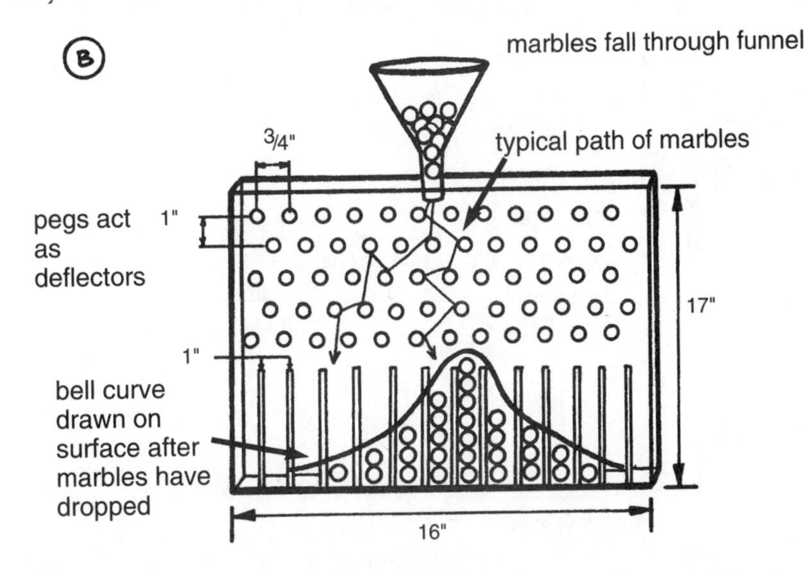

Why?

As you drop the marbles, it's impossible to know what any given marble will do. There are too many collisions with pegs and other marbles. But the marbles fall into the same curve every time (see Half-Life Dating, page 61). Why?

The curve tells you how likely it is that a marble will land in a certain column. The marbles were dropped into the center, so it's more likely that the marbles will end up there. But occasionally, a number of unlikely collisions will send a marble to the edge. The farther from the center you go, the less likely it is that a marble will end up there. That's why fewer balls landed there.

This same curve often describes how test scores are distributed (how many people get what grade). Most students get about the same grade (the bump in the center of the curve). Fewer people get extremely high or extremely low grades (the flattened edges of the curve). Using a type of math called *statistics*, scientists can predict outcomes. While they are not able to predict a particular outcome (such as exactly how many students will get a B), they can predict how likely an outcome is (for instance, that fewer students will probably get a D than a C).

All Wound Up

PARENTAL SUPERVISION RECOMMENDED
Many of the magnets used in factories, laboratories, and homes are not regular magnets. They are *electromagnets*. An electromagnet is created by a current of electricity. Big electromagnets are used in mills and junkyards to move pieces of iron and steel from place to place. Small electromagnets are used in doorbells and in telephones. With this project, you can demonstrate how an electromagnet works.

What You'll Need

- six dry cell batteries (three 6-volt ones and three 9-volt ones)
- small paper clips
- large paper clip
- $4\frac{1}{2}$ yards each of plastic coated (magnetic) wire and plastic insulated (bell) wire.
- box
- wire cutter
- masking or electrical tape
- six nails
- hollow plastic tube (wide enough for a nail to fit inside)

Preparation

1. Cut the bell and magnetic wires into pieces 18", 1 yard, and 2 yards long.
2. Using the wire cutter, take off the plastic coating from all ends of the wires.
3. Open the large paper clip so it looks like a square.
4. Touch this paper clip to each of the nails. If the paper clip sticks, the nails are magnetized. You must find nails that are not magnetized.
5. Wrap the 18" bell wire around one nail so that $1\frac{1}{2}$" remain free on either end of the wire. Do the same with the 1-yard and 2-yard pieces, using two other nails.

Presentation

1. While your viewers watch, tape one end of the coiled 18" wire to the positive (+) terminal of one 6-volt battery and the other end to the negative (-) terminal (A). Now you have turned the nail into an electromagnet.
2. Hang the large paper clip from the nail head as shown (B). Then touch the small paper clips to the large one. They'll stick

because the large paper clip is magnetized. Keep adding small clips until the large clip falls from the nail. How many clips did the large one hold?

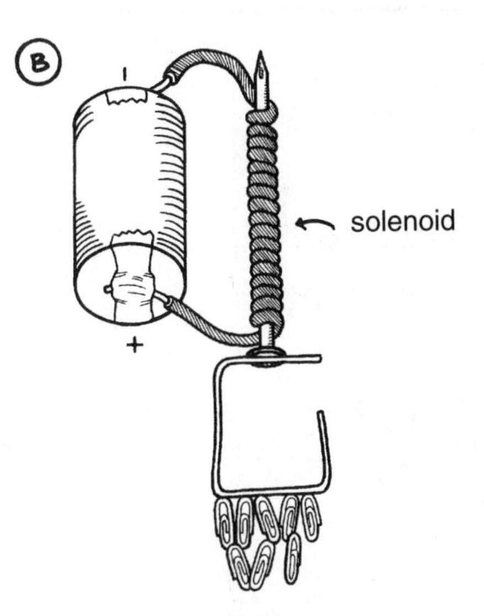

3. Repeat steps 1 and 2 using the 1-yard and 2-yard pieces of bell wire and the other two 6-volt batteries. How many small paper clips can be held by the large clip when using these other electromagnets?
4. Now tell your audience you'd like to repeat this process with three different lengths of magnetic wire to see what happens. Ask for an assistant to help you coil the wire around the three remaining nails, then tape the coils to the three 9-volt batteries. What happens when you repeat step 2 now?
5. Next, with the 2-yard piece of magnetic wire, detach the wire ends from the battery, remove the nail, and replace it with the hollow plastic tube. Then, insert the nail part way into the plastic tube and *very briefly* touch the ends of the wires to the 9-volt battery. The nail will be sucked into the plastic tube! Remove the battery.
6. Reverse the connections to the battery and repeat step 5. Have your viewers predict what will happen.

Why?

Remember that a charge flowing through a wire produces a magnetic field (see Wayward Compasses, page 63)? The field lines surround the wire in expanding circles. When a current-carrying wire is twisted into a coil, called a *solenoid*, the field twists with it, causing the field lines to concentrate inside the coil. This creates a powerful magnetic effect inside the coil—an electromagnet.

Inside a nail are tiny circulating charges that each act as a little magnet with its own magnetic field. These fields usually point in all different directions. But when the nail is inside the coil, the powerful magnetic field in the coil causes the little magnets in the nail to align themselves with the coil's field (north poles all pointing the same way). The little magnets add to the coil field, making it even stronger—strong enough to attract and lift a bunch of paper clips.

Uphill Rollers

Do you think it's possible for an object to defy gravity and roll *uphill*? You can make your audience think so with this project.

What You'll Need

- two yardsticks
- two funnels, 6" in diameter
- stack of books 4" thick
- one book 1" thick
- glue (or tape)

Preparation

1. Glue the large ends of the funnels together.
2. Set up the books and the yardsticks as shown. Make sure the funnels will roll all the way "uphill" (without falling off) when you lay them on the lower end of the yardsticks. If they do not roll uphill, adjust the spacing of the yardsticks at the higher end.

Presentation

1. Ask your viewers if they think the funnels will roll uphill when you place them on the yardsticks.
2. Then put the funnels at the lower end of the yardsticks and watch them roll upward!

Why?

The funnels don't really roll uphill. If you look at them from the side, you'll see that the center of the funnels is really going downhill (B)! Because the sides of the funnel narrow to a small hole, the center of the funnels go down even though the yardsticks slope up.

The center of the funnels rolls downhill.

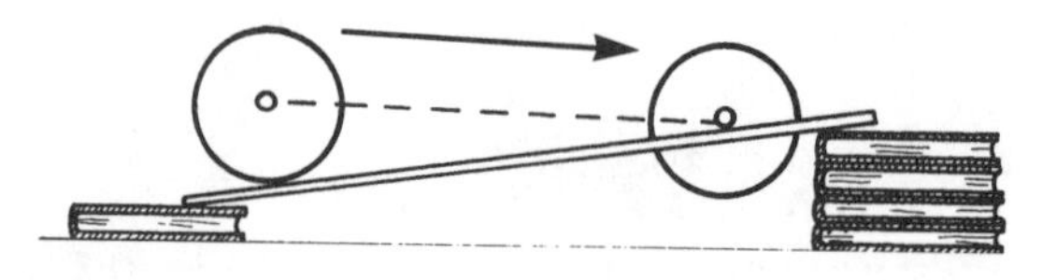

Let 'Em Roll!

Do two objects with the same size and shape roll downhill at the same speed? Set this project "rolling" and see what happens!

What You'll Need

- two identical round, metal cookie tins, with lids
- about 200 pennies
- wooden board, 6' long
- cardboard box, 1' high
- double-sided foam tape

Preparation

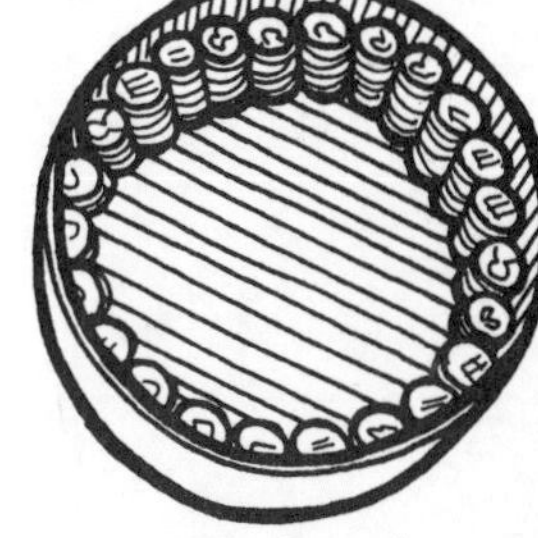

coins placed on edge of tin

1. Tape short stacks of pennies together, using small squares of double-sided tape. Then tape the stacks around the edge of the inside of one cookie tin (A).
2. Next, *using the same number of coins used in step 1*, tape 4 tall stacks together, then tape the stacks to the center of the other tin (B).
3. Put the lids on the tins.
4. Rest one end of the board on top of the box, and rest the other end on a table.

Presentation

1. Ask a viewer to compare the weight of the cookie tins by holding one in each hand. Which tin does he think will roll fastest down the ramp?
2. Put both tins on top of the ramp, and let them go. Did the volunteer guess correctly?

Why?

How is it that two cans of the same size, shape, and weight roll downhill at *different* speeds? When two objects seem the same but behave differently, it means they really *aren't* the same. In the case of the tins, the only difference is that one has coins in the middle of it and one has coins around the edge. This "slight" difference makes all the difference!

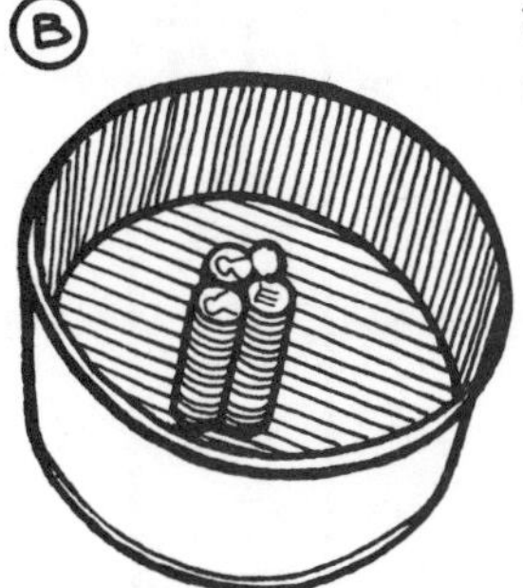

coins stacked in the center

The tin with the coins around the edge rolls more slowly than the other tin because the coins travel around in big circles. By contrast, the coins in the other tin, since they're stacked in the center, move around in much smaller circles. For each revolution (turning) of the tins, the coins on the edge have to move a greater distance than the coins in the center. It takes more energy to get the coins on the edge to move the greater distance. Even though the same force of gravity is pulling both cans downhill, the coins in the center have an advantage—the smaller distance they have to travel means they can speed up faster and win the race.

Let's Clear the Air

You may have heard it said that all objects fall at the same rate. But you know from experience that a feather falls more slowly than a rock. What, then, are scientists telling us? In this project you can "clear the air" and get an answer!

What You'll Need

- Plexiglas™ tube, 4' long, with a 2" to 3" diameter
- two rubber stoppers, large enough to fit snugly in the ends of the tube (one stopper should have a hole in the center)
- copper tubing, 4" long and wide enough to fit tightly through the hole in the stopper (available at hardware stores)
- small fishing weight
- two hose clamps
- flexible plastic tubing, about 6' long
- small hand pump (available at toy stores and auto supply stores)
- petroleum jelly
- small feather

Preparation

1. Put the feather and fishing weight into the tube, then plug up the ends of the tube with the rubber stoppers.
2. Coat the outside of the copper tube with a little petroleum jelly and push it through the hole in the stopper as shown.
3. Slide a hose clamp over one end of the plastic tubing, then put that end over the copper tube. Tighten the hose clamp over the joint so no air can get in.
4. Slide the other hose clamp over the other end of the plastic tubing. Then slip this end of the tubing over the hose on the hand pump. Tighten the hose clamp over this joint, too.

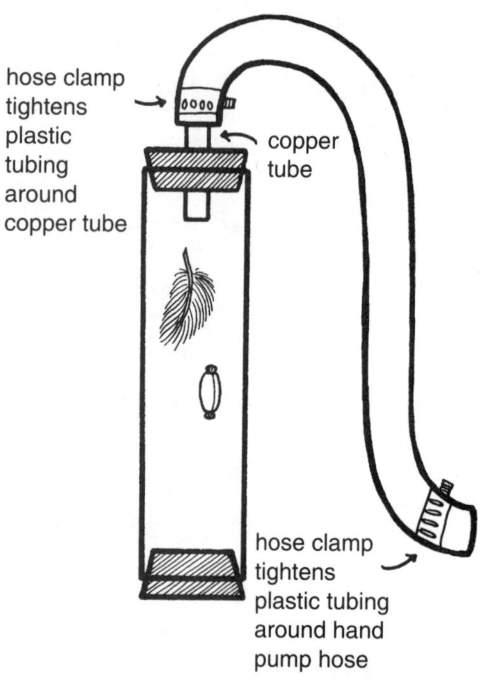

Presentation

1. While your viewers are watching, hold the Plexiglas tube so that the feather and the weight are at one end of the tube. Then quickly flip the tube, letting the feather and weight drop straight down. The weight will hit the bottom first, with the feather falling slowly behind.
2. Now pump the hand pump for a minute or so. Tell your viewers you are creating a vacuum chamber, then repeat step 1. This time, the feather and weight should reach the bottom together (if not, pump some more and try again).

Why?

The weight of an object (how heavy it is) tells you how hard the earth's gravity is pulling it down. So, why does a light feather fall just as fast as a heavy fishing weight when it is pulled down less by the earth's gravity?

The reason is that heavier objects are harder to move. Suppose the fishing weight is 100 times heavier than the feather, so the earth is pulling down on it 100 times harder than on the feather. The heavy weight would also be 100 times harder to move. And while the feather is pulled down by the earth less, it also needs less pull to be moved. These effects cancel each other out and the result is that all objects fall equally. But this idea is only true when there is no air in the way (that is, in a vacuum). As objects fall, they collide with air molecules that slow them down. It is easier for colliding air to slow down a feather than a fishing weight. By pumping, you "cleared the air" out of the way, allowing you to see the effects of gravity alone.

You're the Doctor!

PARENTAL SUPERVISION RECOMMENDED

In the project Listen Up! (page 11), you observed how sound travels through a stethoscope. In "Seeing" is Believing (page 23), you observed how the retina changes when exposed to bright light. The project Hot! Hot! Hot! (page 40) explained how heat causes a material to expand. Did you know that doctors use all these principles to diagnose illnesses in people? In this project, you're the doctor!

What You'll Need

- funnel
- 18"of rubber tubing
- blood pressure gauge and bulb
- glue
- flashlight
- doctor's kit (optional)
- brass paper fastener
- aluminum foil circle, 3" in diameter
- cardboard circle, 3" in diameter
- Velcro™ strip, 24" long
- watch with second hand
- mercury thermometer and its container

Preparation

Stethoscope: See the directions on page 11.

Doctor's mirror:

1. Glue the aluminum foil circle to the cardboard circle.
2. With scissors, cut a ½" hole in the center of the cardboard circle, as well as a small hole near the edge. Also cut a small hole in the middle of the Velcro strip.
3. Attach the cardboard circle to the fuzzy side of the Velcro by pushing the brass paper fastener through the holes in both the cardboard and the Velcro™ (A). You'll wear this mirror during your presentation.

(A)

Thermometer: Make sure the thermometer is properly washed and is clean. Keep it in a safe container.

Presentation

1. Tell viewers you are going to show how doctors use principles of science to diagnose illnesses in people. Announce that you will perform three tests: checking for heart rate, level of consciousness, and blood pressure. Ask for a "patient."
2. Use the stethoscope to listen to the volunteer patient's heartbeat, placing

the funnel over the person's heart (B). Look at your watch and count the number of heartbeats in fifteen seconds. Multiply that figure by four, and announce what your patient's heart rate is. Tell your audience that this example of sound traveling through air allows doctors to detect abnormalities with a person's heartbeat.

3. Next briefly shine the flashlight into the volunteer's eyes. Tell your audience that doctors can determine a person's level of consciousness by seeing if their pupils dilate (grow wider). Explain that if the pupils of an unconscious person do not dilate, the doctor then knows the patient may have suffered an injury to the brain and may be in a coma.
4. Check the volunteer's blood pressure with the blood pressure gauge. Wrap the strap around the person's upper arm. Then pump up the bulb until the strap is tight. What is the person's blood pressure? Explain to your audience that in all the arteries and veins in the body, blood pushes out against the walls of the vessels. When you push on a blood vessel from the outside by using the air pressure gauge, you stop the flow of blood. When the flow stops, that's an indicator of what the blood pressure is pushing out.
5. Now take the person's temperature by putting the thermometer under his or her tongue for two minutes. Explain to your audience that heat causes the mercury to expand, which makes it rise in the thermometer. In this case, the heat is body heat. Since we know what a normal body temperature should be (98.6 degrees Fahrenheit), the thermometer will show us whether or not the person has a fever.

Point of No Return

PARENTAL SUPERVISION RECOMMENDED
Beyond a certain critical temperature called the Curie point (named after Marie and Pierre Curie), the magnetism in matter disappears. Why?

What You'll Need

- cardboard box
- small "donut" magnet
- ruler
- 6-volt lantern battery
- string, about 1' long
- glass of water
- four alligator clips
- two copper wires, 2' long
- iron wire, about 2' long
- oven mitts

Preparation

1. Cut off the top and two sides of the cardboard box.
2. Then cut two notches for the ruler to sit in, as well as a deep "L" shape on either side of the box. Cut two slits for the iron wire to sit in (A).
3. Now tie one end of the string around the ruler and the other end around the magnet. The magnet should just be able to touch the wire.
4. Put two alligator clips on either end of both copper wires.

Presentation

1. In front of your viewers, stick the magnet to the iron wire.
2. Now connect the copper wires to first the iron wire and then to the battery as shown. *WARNING! Because electric current is now flowing through the wire, the wire will get hot! Do NOT touch it.* As the wire becomes hotter, the magnet will eventually fall from the wire.
3. Disconnect the four clip leads, then put on the oven mitts and remove the iron wire. Dip it into the glass of water to cool it down rapidly.
4. Place the iron wire back in the slits and show your viewers that the magnet will stick again.

Why?

When the magnet touches the iron wire, the tiny circulating currents in the wire align themselves (see All Wound Up, page 70) (B). Heating the wire causes its atoms to jiggle about, scrambling the circulating currents and destroying the magnetic attraction of the wire and magnet.

Cosmic Ray Detector

PARENTAL SUPERVISION RECOMMENDED

Every second of every day, invisible high-energy particles shoot through and past you. Where do these particles come from? Most are from the disintegration of radioactive substances, like uranium, here on earth. Others are cosmic rays—particles released into space, perhaps millions of years ago, by exploding stars! By building a cloud chamber, you can show traces of these ghostly particles as they travel through our world.

What You'll Need

- a room in which you can turn off the lights
- thin sponge
- black felt
- rubber cement
- large glass jar with a metal screw-on lid
- plastic insulated gloves
- black construction paper
- isopropyl alcohol
- block of dry ice, 10" x 10" x 2"
- towel (large enough to wrap around the dry ice)
- sheet of cardboard, 1' x 1'
- strong light source (as from a slide or movie projector)
- watch with second hand

Preparation

1. First, cut and glue a circular piece of felt large enough to fit inside the metal lid, as well as a circular piece of sponge large enough to cover the bottom of the jar (A). Let dry.
2. Cut a strip of black paper wide enough to loosely wrap around the jar and a little taller than the jar is high. Glue the strip closed.
3. Cut two slots in the paper cylinder 1" high by 3" wide where shown (A) (the upper one is a viewport, the lower one is for your light source).
4. Fifteen minutes before your presentation, soak the sponge and the felt with alcohol by slowly pouring some in. Then tightly screw on the lid.
5. Put the dry ice on the cardboard and put the jar upside down on top of the ice, making sure the metal lid comes in contact with the ice (B). *WARNING! Dry ice is extremely cold! Wear plastic insulated gloves while handling it. Do NOT touch it with your bare hands.*
6. Wrap the towel around the remaining ice.

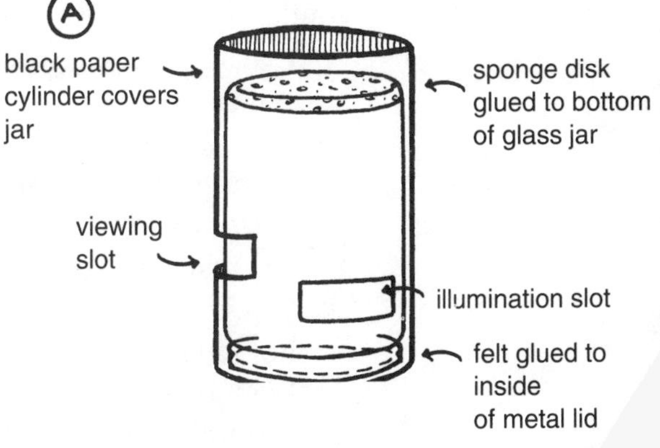

Presentation

1. Slip the black paper cylinder over the jar and arrange the projector light so it shines through the illumination slot as shown. Turn off the room lights.
2. Ask a volunteer to look into the viewing slot toward where the light enters the jar. She will see a miniature rainfall, with tiny threadlike vapor trails appearing out of nowhere. These are the tracks left by high-energy particles as they fly through the jar! Let other spectators peer into the jar to see the particle tracks. Using the watch, count the number of tracks they see in a minute.

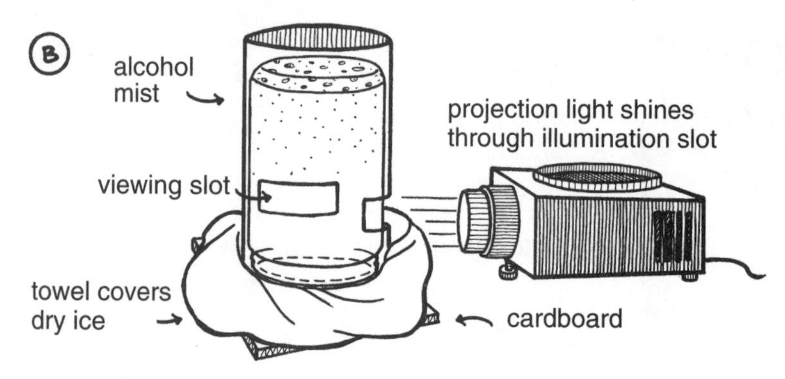

Why?

If conditions are just right, a gaseous vapor (such as water mist) will condense to form liquid droplets. That's what happens in the earth's atmosphere when water molecules condense onto dust particles, form droplets, and fall to the earth as rain.

In your chamber, you created almost all the conditions necessary to make "alcohol rain." As alcohol from the sponge falls to the bottom, it cools and forms a dense, cold gas. All that was needed to form alcohol droplets was something for the alcohol molecules to condense onto (like the dust in the atmosphere that water condenses onto). When an invisible, high-energy particle shoots through the alcohol gas, it collides with alcohol molecules in its path, knocking off some of their electrons (negatively charged particles). Lacking electrons, these alcohol molecules now are positively charged, so they attract the neutral alcohol molecules around them (C) (see Static Magic, page 58). As the alcohol molecules gather, they form small droplets. The trail of droplets is like a trail of footprints marking the path of the invisible particles.

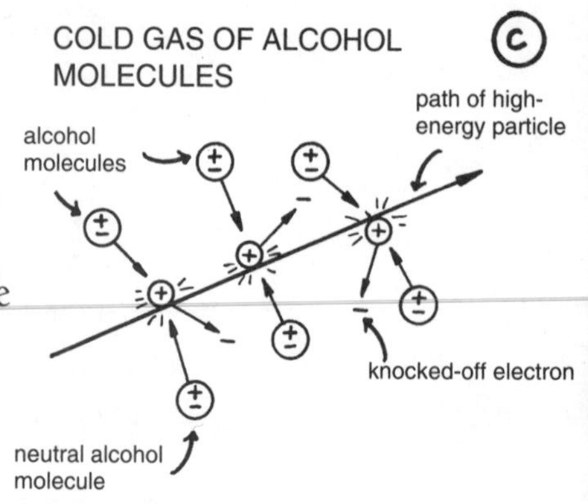